Privatizing War

This book offers a comprehensive moral theory of privatization in war.

It examines the kind of wars that private actors might wage separate from the state and the kind of wars that private actors might wage as functionaries of the state. The first type of war serves to probe the *ad bellum* question of whether private actors can justifiably authorize war, while the second type of war serves to probe the *in bello* question of whether private actors can justifiably participate in war. The cases that drive the analysis are drawn from the rich and complicated history of private military action, stretching back centuries to the Italian city-states whose mercenaries were reviled by Machiavelli. The book also takes up the hypothetical examples conjured by philosophers—the private protective agencies of Robert Nozick's *Anarchy, State, and Utopia*, for example, and the private armies of Thomas More's *Utopia*. The aim of this book is to propose a theory of privatization that retains currency not only in assessing current military engagements but past and future ones as well. In doing so, it also raises a set of important questions about the very enterprise of war.

This book will be of much interest to students of ethics, political philosophy, military studies, international relations, war and conflict studies, and security studies.

William Brand Feldman has a DPhil. in Politics from the University of Oxford and is a resident physician at Brigham and Women's Hospital, a teaching affiliate of Harvard Medical School.

War, Conflict and Ethics
Series Editors: Michael L. Gross
University of Haifa
and
James Pattison
University of Manchester
Founding Editor: Daniel Rothbart
George Mason University

This new book series focuses on the morality of decisions by military and political leaders to engage in violence and the normative underpinnings of military strategy and tactics in the prosecution of the war.

Civilians and Modern War
Armed conflict and the ideology of violence
Edited by Daniel Rothbart, Karina Korostelina and Mohammed Cherkaoui

Ethics, Norms and the Narratives of War
Creating and encountering the enemy other
Pamela Creed

Armed Drones and the Ethics of War
Military virtue in a post-heroic age
Christian Enemark

The Ethics of Nuclear Weapons Dissemination
Moral dilemmas of aspiration, avoidance, and prevention
Thomas E. Doyle, II

Chinese Just War Ethics
Origin, development, and dissent
Edited by Ping-cheung Lo and Sumner B. Twiss

Utilitarianism and the Ethics of War
William H. Shaw

Privatizing War
A moral theory
William Brand Feldman

Privatizing War

A moral theory

William Brand Feldman

Routledge
Taylor & Francis Group

LONDON AND NEW YORK

First published 2016 by Routledge

2 Park Square, Milton Park, Abingdon, Oxon, OX14 4RN

605 Third Avenue, New York, NY 10017

Routledge is an imprint of the Taylor & Francis Group, an informa business

First issued in paperback 2020

British Library Cataloguing-in-Publication Data
A catalogue record for this book is available from the British Library

Library of Congress Cataloging-in-Publication Data
Names: Feldman, William Brand, 1982– author.
Title: Privatizing war : a moral theory / William Brand Feldman.
Description: New York, NY : Routledge, [2016] | Series: War, conflict and
ethics | Includes bibliographical references and index.
Identifiers: LCCN 2016003351| ISBN 9781138803954 (hardback) |
ISBN 9781315753379 (ebook)
Subjects: LCSH: War–Moral and ethical aspects. | Private military
companies–Moral and ethical aspects. | Mercenary troops–Moral and
ethical aspects. | Legitimacy of governments. | Contracting out–Moral and
ethical aspects. | War (Philosophy)
Classification: LCC U22 .F443 2016 | DDC 172/.42–dc23
LC record available at http://lccn.loc.gov/2016003351

ISBN: 978-1-138-80395-4 (hbk)
ISBN: 978-0-367-78728-8 (pbk)

Typeset in Times New Roman
by Wearset Ltd, Boldon, Tyne and Wear

To Mom, Dad, and Jenny

Contents

Acknowledgments

This book was made possible by financial support from the United Kingdom Overseas Research Students Awards Scheme, Nuffield College, the University of Oxford Department of Politics and International Relations, and the Harry Frank Guggenheim Foundation. I would also like to thank Yale University for hosting me as an Exchange Scholar from 2008–10.

Four political theorists greatly influenced this project. Henry Shue helped me to formulate its initial questions and provided valuable feedback on early chapters. David Miller commented on the initial project proposal. Both Henry and David have been an inspiration to me, for the quality of their scholarship and the patience and generosity with which they advise students. Steven Smith gave me the opportunity to serve as his teaching assistant and then became an advisor away-from-home and good friend. Most importantly, Simon Caney has been a source of personal support, incisive criticism, and inestimable scholarly knowledge. He improved my arguments with an ever-present humility and gentleness of spirit that I seek to emulate in my relationships with others. Those who know his work will spot his influence, both methodological and substantive, at every turn in this book.

I also owe a debt of gratitude to many friends who discussed military privatization with me during the past several years: Jeremy Farris, Ben Gross, Carolyn Haggis, Sylvia Houghteling, Jeremy Kessler, Fred Meiton, Tom Schmidt, Matt Shapiro, Ezra Siller, Keith Stanski, and Gabriel Wolner. A special thanks to Ashwini Vasanthakumar, who taught me a great deal about political theory at Blue State Coffee. James Pattison and Daniel McDermott offered sharp commentary during my *viva voce*, where much of the material in this book was initially presented. Thank you to two anonymous reviewers from Routledge and the editors of the War, Conflict and Ethics series who provided valuable comments on the book. The library staff of Nuffield College, Yale University, and the University of California, San Francisco have been superb.

Lastly, I would like to thank my family. My father, Stuart Feldman, first introduced me to political theory when I was seven in a memorable conversation over hot chocolate in Minneapolis. His unwavering commitment to my success has been a constant source of comfort. My mother, Johanna Feldman, balanced her responsibilities as a parent with her career as a prosecutor in a way that I could only dream of doing myself one day. Her compassion and humanity have allowed me to pursue my interests. For that, and much else, I will forever be grateful. My sister, Jenny Feldman Eskildsen, continues to awe me with her outlook on life and her medical savvy. My fiancée, Melissa Danesh, remains my closest friend, favorite S-room study partner, and steadiest guide.

The mistakes in this book are all mine.

1 Introduction

1.1 The Middleman state

In 2013, the United States (US) federal government employed an estimated 11.6 million people: 2.1 million civil servants, 500,000 postal workers, 1.4 million active duty military personnel, and 7.6 million contractors (Schwellenbach 2014).[1] These figures tell a number of intriguing stories, but probably none is more conspicuous than the magnitude of government outsourcing. Only one-third of federal jobholders—the four million civil servants, postal workers, and military personnel who constitute the first three groups—were *directly* employed by the government. The remaining seven and half million, a so-called "shadow government," were contractors who worked alongside federal officials in nearly every government sector and performed many of the tasks that they performed, ranging from tax collection and budget preparation to software development and cattle husbandry, but who were not themselves government officials.

Reliance upon these shadow employees reflects a simple political philosophy that has come to guide Washington: "the business of government is not to provide services, but to see to it that [services] are provided" (United States House Committee on Government Reform 2006: 1).[2] President Clinton called it "reinventing government" and President Bush "the managerial revolution" (Frank 2008: 134–135).[3] President Obama, though more critical of outsourcing— the "massive cost overruns, outright fraud, and the absence of oversight and accountability," as he puts it (Zeleny 2009)—has aimed to reduce its wastefulness (Moorhead 2012; Ivory 2014)[4] but not scrap his predecessors' core commitments.[5]

Well-regulated private companies, the thinking goes, inject efficiency into a notoriously unmanageable system and yet still provide citizens with the same, if not better, services, driving governmental operating costs down without jeopardizing individual entitlements. On the one hand, this approach eschews what has been dubbed the Nanny State, a traditionally ambitious mixture of bureaucratic largesse and expansive policy agendas. Yet, on the other hand, it refuses to endorse the kind of wholesale minimalism that we might associate with the Nanny State's foil, the so-called Night Watchman State, which is defended by some libertarians. The new model straddles these two prototypes, combining small bureaucracies with far-reaching policy agendas. The result, we might say, is a Middleman State, one that itself furnishes little but through others procures a great deal.

The realm where Middleman tendencies are most pervasive is the military. The US Department of Defense (DoD) spends more money on contracts than every other federal agency *combined* (Schwartz and Ginsberg 2013). DoD contracts represent approximately 70 percent of all federal government contracts and 10 percent of the entire US federal budget (Schwartz and Ginsberg 2013). The US government, we might say, has largely functioned in recent years to furnish a capacity for war, and it has furnished this capacity with a military that reflects Middleman ideology.

1.2 The boundaries of procurement

Many military functions, of course, are carried out away from the battlefield: on munitions factory floors, for example, or in shipyards where a disproportionate ratio of contractors to regular personnel would be expected. Militaries throughout history have relied upon the private sector for *some* support, if only to grow the food its soldiers eat or sew the uniforms they wear, and a number of outsourced services, like munitions production and ship construction, may be hard to distinguish from tasks that fulfill these more inescapable dependencies.

But battlefields have come to display the ballooning ratios of contractors to regular military personnel that we see on the whole. In Iraq and Afghanistan, during the peak of hostilities, more than one contractor served for every one member of the military, up from one in 25 during the Gulf War (Schwartz and Swain 2011).[6] As the US attempts to wind down its war in Afghanistan (at the time of this writing), 39,000 contractors serve on the ground alongside 12,000 regular soldiers (Shorrock 2015). Generally, contractors perform a range of functions. They drive truck routes, clean latrines, perform interrogations, cook meals, train police, wash clothes, repair vehicles, build bases, protect diplomats, translate interviews, and gather intelligence (Pan 2004; Fainaru 2007; Isenberg 2009). Sometimes, they also engage in killing. Employees of Blackwater, for instance, one of the more than 300 private security firms that operated in Iraq, resorted to an "escalation of force," which is to say a discharge of weapons, on 195 separate occasions from 2005 to 2007 (Glanz 2008; DeYoung 2007). Though much of the killing by Blackwater and other firms remains shrouded in secrecy, reports paint a picture of contractors engaged in combat—of decisively militarized Middleman functionaries.[7]

The simple puzzle is where to draw lines. When are Middleman practices no longer permissible? Perhaps a boundary should be drawn around the entire battlefield, or rather where the sound of gunfire can be heard.[8] Perhaps the appropriate restriction is not on physical proximity to combat but on participation in combat or not on participation in combat but on participation in "offensive" (as opposed to "defensive") combat (Runzo 2008: 68).[9] Some have even questioned whether lines should be drawn at all (Fabre 2012; Baker 2011a). Is there a morally relevant distinction between hiring private ground controllers and renting an air force or between defending an installation and driving away its assailants? What if the Middleman State were simply to procure its entire military from the private sector? We see no problem when the Middleman State asks the private sector to provide its citizens with newly constructed roads. Why should it behave differently on the battlefield?

Beneath these questions about military combat is a still more heretical set of questions about the very enterprise of war. Why, we might ask, should states be involved in war in the first place? What if the Middleman State were cut out entirely and the initiation of war left to the private sector? In one of the more disturbing manifestations of private military action in the twenty-first century, a group of entrepreneurs from Britain and South Africa hatched a plot in 2004 to overthrow the ruling power of Equatorial Guinea for the purpose of securing oil.[10] The planned overthrow bizarrely mirrored a fictionalized coup depicted by Frederick Forsyth (1974) in his novel *The Dogs of War*, in which a group of business operatives overthrow an African dictator (Polgreen 2008a, 2008b). Ultimately, the real coup was derailed before being launched: as Simon Mann, the coup's mastermind, and 70 others prepared to board a plane stocked with weapons headed for Equatorial Guinea, they were arrested at the Harare airport.[11] But the plans provide a provocative glimpse into an application of private military force that goes beyond the outsourcing upon which US forces have relied in Iraq and Afghanistan, pushing the debate past the mere limits of the Middleman State to include the prospect of fully privatized war.

1.3 The project and argument

The book that follows takes up this debate. The aim is to deliver a comprehensive moral theory of privatization in war: that is, to delineate the *full* spectrum of justifiable military action for private agents. In service of this goal, I ask about the kind of wars that private actors might wage apart from the state, as Mann hoped to wage in Equatorial Guinea, and about the kind of wars that private actors might wage as functionaries of the state, as firms like Blackwater waged in Iraq.

The first set of wars, we might say, serves to probe the question of military *authorization* and the second set of military *supply*. This distinction—between authorization and supply—is at the center of this book and so merits a brief elaboration. The provision of any good, including military force, consists in these two discrete undertakings.[12] The entity that undertakes authorization decides both that military force will be supplied and by whom. The entity that undertakes supply then executes the various tasks that have been authorized for performance. The same entity, of course, may assume these roles together, as the US largely did in the Gulf War (and most previous wars), but the roles may also be divorced, as we have seen with US action in Iraq and Afghanistan. If an account of privatization in war is to be rigorous and capable of providing guidance in the world that we face, it must be carefully attuned to the myriad ways that a government might involve or fail to involve itself in the use of military force. As such, authorization and supply will each be treated in detail.[13]

The cases that drive my analysis are largely drawn from the rich and complicated history of private military action, stretching back centuries to the Italian city-states whose mercenaries were reviled by Machiavelli. Occasionally, I examine the hypothetical examples conjured by philosophers—the private protective agencies of Robert Nozick's *Anarchy, State, and Utopia*, for example, and the private armies of Thomas More's *Utopia*. All is with an eye toward fleshing out a theory of privatization that retains currency not only in assessing current military engagements but past and future ones as well. The width of this investigation means that a multitude of actors with an array of names, historical and otherwise, fall into its purview: *condotierre*, freedom fighters, guerillas, insurgents, joint-stock companies, *landsknechts*, mercenaries, militias, pirates, private military companies, private protective agencies, private security companies, privateers, revolutionaries, and terrorists, to name but a few. What they all have in common is their participation in the private provision of war, whether in its authorization, supply, or both.

This book advances two central lines of argument. (I) Public entities have a right, and indeed a duty, to monopolize the authorization of military force. Private authorization, in the mold of Simon Mann's plot on Equatorial Guinea, is morally unjustifiable. Only public military authorization, I argue, is justifiable. (II) Once military force has been publicly authorized, authorizing entities have a further duty to supply military force under the direction of public military leaders (defined as commanders and their superiors on the chain of command). By contrast, public entities may hire private agents to discharge the responsibilities of rank-and-file personnel (those who serve below commanders on the chain of command) so long as they place these personnel within a uniform chain of command. Groups like Blackwater therefore may engage in the supply of military force on behalf of public entities if they serve under the command of these entities. Part I will argue for the first set of claims and Part II for the second set.

Before turning to these claims, I want to briefly sketch four sets of preliminary remarks that should help clarify the chapters to come. The remainder of this chapter presents these remarks. First, I provide working definitions for two key terms of the book (and its title): "war" and "privatization." Second, I explain in further detail why a book on privatizing war is valuable. Next, I defend a framework for analyzing the privatization of war. Finally, I outline how each chapter fits into the framework for analysis that I propose.

1.4 "War" and "privatization"

Let us begin with the meaning of "war." The conception that I wish to defend largely follows the conception that C.A.J. Coady (2008: 4–8) offers. According to Coady, "war is the resort by an organized group to a relatively large-scale act of violence for political purposes to compel an enemy to do the group's will."

The only amendment that I want to suggest is that the phrase "for political purposes" be removed. The notion that wars are necessarily political, which is extracted from Karl von Clausewitz's (2008 [1832]: 28) famous dictum that war is the "mere continuation of policy by other means," is not a description of what war *is* but rather a prescription for what it *ought* to be.[14] Wars, we must acknowledge, have been fought for a multitude of reasons, some of which are purely financial and many of which, to make a more general point, are a function of the drives and even the culture that constitute human nature. As John Keegan (1994: 3) eloquently puts this latter point, warfare

> is almost as old as man himself, and reaches into the most secret places of the human heart, places where self dissolves rational purpose, where pride reigns, where emotion is paramount, where instinct is king. "Man is a political animal," said Aristotle. Clausewitz, a child of Aristotle, went no further than to say that a political animal is a war making animal. Neither dared confront the thought that man is a thinking animal in whom the intellect directs the urge to hunt and the ability to kill.

The mistake that Coady makes in following Clausewitz and Aristotle risks not only foreclosing potentially instructive lines of moral thought but also jeopardizing the rules that we endorse by smuggling in assumptions that have not been defended. If our aim is to reduce the hellishness of war through reason and argumentation, we must try to reason and argue about all mechanisms of constraint, including the institutional constraint of relegating war to the political realm.

But the remainder of Coady's language, which seems plausible, will be preserved. "Organized groups" rather than "states" are kept as the subjects of analysis. Any rendering to the contrary would exclude a host of conflicts that we typically recognize as wars: revolutionary and secessionist wars like the American Revolution and Civil War, client wars like the joint US–Mujahideen action against the Soviet Union in Afghanistan, tribal wars like Native American resistance to European colonization, genocidal wars like the Hutu slaughter of Tutsis in Rwanda, terrorist attacks like the bombings of September 11 on the Pentagon and World Trade Center, and the subsequent US-led "war on terror" aimed largely at non-state actors.[15] Moreover, in using the phrase "organized groups," we also capture the exclusion of disorganized violence committed by isolated actors; wars are coordinated efforts, not spontaneous outbursts. Finally, the phrase "relatively large scale" is preserved to qualify the act of violence since, as Coady rightly points out, wars are fought on a scale that exceeds conflicts like criminal conspiracies, feuds, riots, assassinations, disturbances, skirmishes, fomentations, plots, fracases, scuffles, gang fights, brawls, and clashes. Any attempt to cleanly quantify the ingredients of magnitude that are sufficient to motivate the label "war," say by counting the number of dead, gunshots fired, or the duration of hostilities, will surely ring false.[16] We must be prepared to debate specific cases and acknowledge that our judgments about these cases may sometimes diverge. But wars are elaborate endeavors, which may be examined apart from more circumscribed cases of violence that Michael Walzer (2006a: 106) helpfully labels "force-short-of-war."[17]

To be clear, this is not to endorse a bifurcated morality that applies one set of standards to war and a different set of standards to activities other than war. I share the sentiment of many contemporary political theorists who defend a strict continuity between the morality of war and the morality of everyday life—a morality that nevertheless must be unstintingly cognizant of the profound deviance that war represents, of its exigencies and of its savagery.[18] Rather, the application of a ceiling below which conflicts constitute force-short-of-war simply enables a faithfulness to everyday language and a pragmatism about the limits of what a single philosophical work can accomplish. That said, since some conflicts are harder to characterize than others, I analyze straightforward instances of war to the extent possible and restrict my conclusions accordingly.[19]

As for the concept of "privatization," it will connote, simply speaking, a trend away from publicly provided goods toward privately provided goods.[20] According to the distinction that I will employ between publicly and privately provided goods (which is admittedly crude), the former are goods that are provided by governments, and the latter are goods that are provided by entities other than governments. As with so many distinctions in political theory, challenging examples fit poorly into both categories. A guerilla movement, for instance, that develops government-like political mechanisms for decision-making but is not itself a governmental entity may strike us as public rather than private. So too might its soldiers if they fight to advance the movement's cause. Conversely, governmental leaders who authorize military force in exclusive furtherance of their own business interests, say, oil executives who take control of a government to fight wars to expand drilling sources, may more closely resemble private actors engaged in authorization than public actors. And soldiers who are supplied to a war effort in exchange for large payments to a government's leadership—as, for example, the "Hessians" were supplied during the American Revolution by Frederick II in exchange for vast payments from the British—may be better categorized as private than public.[21] For now, I will ignore these complexities; the public–private distinction, understood simply as a distinction between governmental and non-governmental provision, should be sufficiently crisp to satisfy the needs of my analysis until an account of who may justifiably exercise military force is better elaborated.

To sum up then, this book investigates whether and to what extent non-governmental actors, as part of some organized group, may participate in relatively large-scale resorts to violence that compel their enemies to the group's will, whether these actors participate by deciding that the resort to violence will be undertaken and by whom (authorization) or by undertaking the various tasks that have been slated for performance (supply). All moves that deviate from the public provision of military force—that is, from public authorization and public supply—will be assessed and thus each stage of a shift from public provision towards private provision analyzed.[22] The yield is a moral account of *privatized war*.

1.5 Why examine privatized war?

A comprehensive work on privatized war is needed for several reasons. The first broad value of this book is that it helps answer two questions, one at the heart of political theory and one that should be wrenched from its periphery: (*a*) what are the goods that a government ought to provide its citizens and (*b*) which of these must be furnished by the government itself? Governments around the world guarantee their citizens a varying range of provisions, including military force, law enforcement, education, welfare, health care, roads, economic regulation, waste disposal, and postal delivery. The provision of some of these goods is almost inseparable from what it means to govern (would we even use the term "government" for an entity that did not enforce any laws?). Others could more plausibly be defended as discretionary. We may not chastise a government that abstained from collecting its citizens' garbage or delivering their mail in the way that we would condemn a government that chose not to educate its citizens or feed those in poverty. Whether governments have a responsibility to ensure the provision of certain goods, and not others, depends upon the value of those goods and upon the appropriate role of government in securing values of that sort. In questioning whether the provision of military force is an essential governmental responsibility, this book seeks to pin down the role of government in one important sphere.

And in asking whether and to what extent, in the military sphere, a government must furnish the good itself, this book, instead of neglecting perplexing questions about the full nature of public provision in a given realm, attempts to develop it. Just as the public responsibility to ensure that certain goods are provided varies depending upon the nature of the good in question, so too will the duty for a government to furnish the good itself. We may not balk at the practice of outsourcing garbage collection, but we would likely recoil at the prospect of the government outsourcing criminal justice, say by hiring private companies to decide court cases and issue prison sentences. Is the provision of military force more like mail delivery or more like criminal adjudication? Though this book cannot speak to outsourcing in all spheres, it will arrive at a conclusion about outsourcing in one important, if commonly public, sphere of activity.

While the ambition of determining the appropriate role of government in the authorization and supply of only one good, military force, may seem modest, there is no particularly compelling alternative. To know whether the government has an obligation or right to provide a good will depend upon the nature of that good, and the nature of a good like military protection is very different from the nature of a good like garbage collection (still further from goods, not yet mentioned but often provided publicly, like arts sponsorship and space exploration). Rather than concoct abstract principles to generate a list of every good that a government may or must provide, we are better off examining the full particularities of a single good and then turning to others. Privatization poses a wealth of complications for war, hardly capable of being fully addressed even in a manuscript of this length (perhaps other goods will require a treatment far less extensive). Only a piecemeal approach can, I think, pay sufficiently close attention to the uniqueness of war while still theorizing with substance.[23]

That said, the contribution of this book to the question of public provision in general goes beyond the specific slice that I carve out and analyze. Its structure delivers a method for analyzing the provision of any good that a government might undertake. As we will see, the tactic that I chart is to identify why a government should see to it that a good is provided (authorization), be it military force or any other provision, and then analyze whether such a rationale coheres with a logic of outsourcing (supply). This strategy will, I hope, serve as a model for future research on parallel questions concerning other goods. The book also articulates a language and set of concepts that, in conjunction with the methodology that I present, may further facilitate investigation of similar topics. Finally, since and insofar as some of our reasons for valuing military provision match our reasons for valuing the provision of other goods, the substantive conclusions that I defend regarding war may, in some instances, be applied to other goods. Where relevant, I draw the reader's attention to these implications.

In addition to the value generated by this book for understanding one realm of public provision (be it any realm) and, in so doing, advancing a more general account of public provision, the second broad value of this book is a product of the fact that the practice of contemporary war is *particularly* desperate for a theory of privatization. As scholars and practitioners begin to reflect with some remove about what transpired in Iraq and Afghanistan and how we are to learn from these troubled invasions and occupations, the topic of privatization will (and must) be central to these reflections. As the wars and indeed the many Middleman examples that will be explored indicate, the moral injunction against military privatization has eroded. And yet despite the crumbling, perhaps unsustainable boundaries that we tend to draw between core military functions and peripheral military functions, a norm against wholesale privatization remains. The idea that states would hire men and women from outside of their own armed forces to pursue and kill enemies in war continues to repel many. To see this, we need only look at the market shift away from purportedly combat-oriented private military companies (PMCs) to support-oriented private security companies (PSCs).[24] As Sarah Percy (2007a: 227) shrewdly argues, we could not make sense of this shift without a norm against some forms of privatized combat, for in the absence of such a norm, companies and states would have no obvious reason to emphasize the defensive, non-combative nature of their operations.

Yet, these norms are neither specific nor obviously correct (perhaps they are too weak, perhaps too strong). The lack of specificity results in a general confusion about what our norms actually require. The possibility that these norms lack a satisfactory *moral* grounding leaves us hesitant about whether we should even care what these norms require. And the increased prevalence of activity that might contradict these norms makes us (or at least should make us) anxious to find the specificity and justification that is missing. If, as I believe to be the case, one of the goals of political theory is to question *how* and *whether* unspecified or poorly specified norms in society should be held, particularly when practices seem to be deviating from our broad, even if vague, instincts about what the norms demand, then a philosophical examination of privatization in war is essential.

Despite such a need, contemporary theorists who study war have only begun to address the moral implications of privatization. Several have examined the just war requirement of "legitimate authority" in the initiation of war.[25] Others have analyzed the permissibility of military outsourcing.[26] But few have developed an account of military authorization *in conjunction with* an account of military supply to offer a comprehensive theory of military privatization. This disjoint has resulted in a failure by contemporary theorists to consider how the principles that govern the former might constrain the latter (and vice versa).

The two exceptions to this divided coverage are Cécile Fabre's (2012) *Cosmopolitan War* and James Pattison's (2014) *The Morality of Private War*. Fabre advances a *cosmopolitan* defense of private authorization and private supply, while Pattison develops what he calls a *cumulative legitimacy* argument for (mostly) public authorization and public supply. Both Fabre and Pattison provide philosophically rich and expansive treatments of military privatization. But neither, I will argue, develops adequate constraints on private actors in war. Fabre fails to consider how even unflinching cosmopolitans might object to privatization, and Pattison, while capturing these objections, does not successfully establish the first principles upon which they rely. This book will aim to fill these scholarly gaps. Hopefully, in so doing, it will say something of interest to the political leaders, military commanders, and even the informed citizens and soldiers who confront these moral questions.

1.6 Societal arrangements for the provision of military force

Having explained the concepts of war and privatization and given a rationale for this inquiry, I turn now to the framework that will be used to structure my analysis. Recalling the two sets of distinctions above, between authorization and supply and between public and private provision, there are four distinct variants for the provision of military force, or indeed for the provision of any good, that one can specify. First is what we may call the *pure private variant*, which entails both private authorization and private supply: the war making of the Dutch East India Company, for example, which ultimately amassed a private army of more than 100,000 troops and enjoyed a Charter that, according to the *Universal Dictionary* of 1751, thrived because it possessed "a kind of sovereignty and dominion" and could make "peace and war at pleasure, and by its own authority" (Singer 2007: 34). The second variant, what we may call the *privately outsourced* variant, entails private authorization but public supply.[27] This variant is well-illustrated by the decision of British Petroleum (BP) to hire a battalion of Colombian soldiers from the government for more than $50 million to protect its oil installations (see Schemo 1996; Singer 2007: 14).[28] The third variant, the *publicly outsourced variant*, entails public authorization but private supply. The contractors stationed in Iraq and Afghanistan, discussed above, fall into this category. Finally, the *pure public variant*, for which an abundance of combat operations in most wars during the last several centuries are representative, entails both public authorization and public supply.[29]

Of course, when we ask about the provision of some good *g*, it may be the case that all of *g* is provided according to one variant or that *g* is provided according to several variants. We could imagine the provision of military force in a given society, for example, entirely according to the public variant, if the government were not to contract any military functions to the private sector, or entirely according to the publicly outsourced variant, if the government were to contract *every* military function to the private sector. We could also imagine a combination of the two, if the government were to authorize military force and supply certain military functions itself while hiring contractors for others, as the US now does—or a tripartite combination, if the US were to stand aside as private groups authorized wars of their own (the prospect of humanitarian action by groups like Blackwater may not be so farfetched).[30]

Generally, for any good *g*, whatever *g* might be, including military force, provision may be undertaken according to one of the four variants or some combination of the four (or, of course, not at all).[31] While different sets of combinations (or *arrangements*) may be of special interest in the investigation of different goods, six arrangements are particularly important for the investigation of war.

The first arrangement (A_0) is the *non-provision of military force*, which describes the state of affairs envisioned by some pacifist theories of war in which *no* military force is provided. Second is what we may call a *free market economy of military force*, usually defended by anarcho-libertarians, in which individuals privately authorize military force to the full exclusion of governmental authorization (A_1). Such a model is embodied by the state-of-nature protective associations that philosophers like Nozick (1974: 12–17) have imagined but also by a real-world example beloved by anarcho-libertarians: the Irish *tuath*, a purely voluntary association for the settlement of disputes and the authorization of war in medieval Ireland.[32] Next is a *mixed market economy of military force* where the government functions as one agent who authorizes war but alongside private agents who do the same (A_2). In those areas where the Dutch East India Company provided military force for some while local governments provided military force for others, a mixed market economy was operative. The fourth arrangement, what we may call an *outsourced public monopoly over military force*, involves governmental authorization but fully private supply—the possession of a military, in other words, but one that is fully leased (A_3). Such a military was the force of choice for Thomas More's (2002 [1516]: 88) Utopians, whose citizens hired "mercenary soldiers from everywhere," because they held "their own people dear, and value[d] one another so highly that they would not willingly exchange one of themselves for an enemy's prince." The fifth arrangement, which many countries now have, is a *partially outsourced public monopoly over military force*, under which the government hires out some of its war making efforts to the private sector (A_4). Finally, *a comprehensive public monopoly over military force* entails both governmental authorization and governmental supply; it is the model towards which the international community seemed to long aspire (A_5).[33]

With this schematic in mind, the concept of *privatization* may now be elaborated with further clarity.[34] Privatization refers to the downward trend from higher-level arrangements to lower-level arrangements on the spectrum A_1–A_5 and, within each arrangement, to the trend away from public provision of military force towards privately outsourced, publicly outsourced, and ultimately private provision.[35] Conversely, *nationalization* refers to the upward trend from lower-level arrangements to higher-level arrangements on the spectrum A_1–A_5, and, within each arrangement, to the trend away from private provision of military force towards privately outsourced, publicly outsourced and ultimately public provision.[36] Military force may be thought fully privatized when private entities enjoy a complete absence of government in a free market economy of military force and fully nationalized when some public entity, be it the state or a sub-state or supranational organization, enjoys a comprehensive monopoly over the provision of military force.[37]

Table 1.1 Different possible arrangements (A_n) for the provision of military force in a society.

Arrangement	Variants included	Entailment
A_0 Non-provision of military force	No variant	*No* military authorization or supply
A_1 Free market economy of military force	Private variant	*Exclusively* private military authorization *Exclusively* private military supply
A_2 Mixed market economy of military force	Combination of private, privately outsourced, publicly outsourced, and public variants other than the combinations enshrined in A_0–A_1 and A_3–A_5	Private *and* public military authorization Private *and/or* public military supply
A_3 Outsourced public monopoly over military force	Publicly outsourced variant	*Exclusively* public military authorization *Exclusively* private military supply
A_4 Partially outsourced public monopoly over military force	Publicly outsourced and public variants	*Exclusively* public military authorization Private *and* public military supply
A_5 Comprehensive public monopoly over military force	Public variant	*Exclusively* public military authorization *Exclusively* public military supply

1.7 The chapters

With a framework for the societal provision of military force in mind, let us finally turn to the specific structure of this book. Part I (Chapters 2–4) argues that private military authorization is unjustifiable and that public military authorization alone is justifiable. Chapter 2 frames my argument within the existing literature on legitimate authority and challenges an important defense of private military authorization (that of Cécile Fabre). In Chapters 3 and 4, I present what is dubbed the *risk-prevention argument*. Public actors, I seek to show, have a duty to prevent private actors from authorizing military force (public *monopolization*). I defend two premises to support this position. The first, which I call the *all affected fundamental interests premise*, is defended in Chapter 3. This premise claims that a decision ought to be withdrawn from the private sector and reserved for public discharge when one course of action under consideration imposes risk, above some threshold level, to the fundamental interests (e.g., physical security) of enough individuals. Fundamental interests are defined as those interests that are sufficiently weighty so as to be protected by rights. The second premise, dubbed the *risk-prevention of war premise*, is defended in Chapter 4. This premise maintains that the private decision to authorize military force *does* impose risk, above an acceptable threshold level, to the fundamental interests (particularly in physical security) of enough individuals. From these two premises, we may conclude that the decision to authorize military force ought to be withdrawn from the private sector and reserved for public discharge. Military authorization, in other words, should be publicly monopolized and A_1 and A_2 rejected.

Once this theory of public *monopolization* is finalized, I then offer a theory of public military *authorization*. The need for a defense of public military authorization *after* a defense of public monopolization may seem nonsensical. For some, public actors who monopolize the authorization of military force would, by definition, *both* prevent private actors from authorizing military force and authorize military force themselves. But this book understands the term "monopolization over military authorization" in a different fashion. On my account, public actors who monopolize the authorization of military force prevent private actors from authorizing military force but without *necessarily* authorizing military force themselves. Because public actors can, in theory, bar private actors from authorizing military force using methods other than war—for instance, intelligence gathering, policing, and imprisonment—a defense of public military monopolization, understood in the second sense, need not constitute a defense of public military authorization. While linguistic convention might favor the first understanding, I will adopt the second. It is more suitable to my analysis and not so divorced from ordinary language as to be unrecognizable. The *public monopolization of military authorization* then, to be precise, will refer to the withdrawal of decisions of military authorization from the private sector and the reservation of these decisions for public discharge.

We should observe that the public monopolization of military authorization, defined in this way, is consistent with some pacifist theories of war, which advocate the government-enforced prevention of private military authorization but without government authorization of war. It is also consistent with just war theories that permit public military authorization in conjunction with the government-enforced prevention of private military authorization. Whether public entities that monopolize the authorization of military force ought to enshrine A_0 or A_3–A_5 will simply depend upon whether the public authorization of military force can be justified.

Thus, after finalizing the risk-prevention argument in Chapter 4, I consider whether public monopolization ought to be accompanied by a prohibition on public military authorization or instead by an endorsement of public military authorization. I argue that, once military force has been publicly monopolized, governments have a further obligation to protect individuals who have been disarmed. To do so, governments must sometimes authorize military force on their behalf. This argument, referred to as the *compensation argument*, rules out A_0.[38]

Having completed rejections of A_0, A_1, and A_2 in Part I, I shift my attention in Part II (Chapters 5–8) to A_3–A_5. What distinguishes these arrangements is the place that each affords to private military supply—whether an exclusive place (A_3), some place (A_4), or no place (A_5). In Chapters 5 and 6, I argue that *some* military supply must be public. First, in Chapter 5, I present what is dubbed the *governance argument* to show that militaries may not justifiably outsource the responsibilities of *high-ranking military officers* to the private sector. These officers, I argue, exert political power over civilian decisions of military authorization and supply. Moreover, they make weighty strategic and tactical decisions *themselves* on behalf of a citizenry in battle. For each of these reasons, the decision-making responsibilities of high-ranking military officers ought to be publicly discharged. In practice, these high-ranking personnel might include flag grade officers (generals) and field grade officers (colonels and majors). They are what Colin Powell calls "the action officers connecting the military forces to the political system and the political system back to the forces" (Bland 1999: 15).[39] What defines these individuals, however, is not the rank that they hold but the extent to which they contribute to civilian decisions of military authorization and supply and enjoy wide discretion in making strategic and tactical decisions in battle.

In Chapter 6, I supplement the governance argument with what I call the *punishment argument*. The punishment argument aims to show that not only must governments refrain from outsourcing the responsibilities of high-ranking military officers to the private sector, but they must also refrain from outsourcing the responsibilities of commanders to the private sector. Commanders are typically the company grade and non-commissioned officers who serve below high-ranking military officers (flag grade and field grade officers) but above rank-and-file personnel.[40] I seek to show that commanders must possess a responsibility to punish disobedient rank-and-file personnel—sometimes with methods that severely restrict individual freedoms (like imprisonment)—and that these methods ought to be reserved exclusively for public discharge. Taken in conjunction, the governance argument and the punishment argument provide an account of functions in military supply that must be publicly performed. Neither the responsibilities of commanders nor their superiors, high-ranking military officers, may be outsourced. In short, the *military leadership*—a term that I use to identify high-ranking military officers and commanders together—ought to remain public.

Chapter 7 argues further that a government may privatize those responsibilities that are discharged by *rank-and-file personnel*, taken to be the set of individuals who serve outside the military's leadership, *only if* such individuals are placed under the military's uniform chain of command. Finally, Chapter 8 considers and dismisses a number of objections to private military supply.

With the structure and primary concepts that animate this book in place, we may now turn to Part I and consider private military authorization.

Notes

1 The number of contractors in 2013 was an estimate based on Paul Light's (2006) analysis in 2005, which relied upon the United States Federal Procurement Data System and the Federal Assistance Data System. No similar estimates have been undertaken since 2006. Schwellenbach (2014) assumes that the number of contractors remained constant from 2005 to 2013 since the value of contracts awarded during this period remained largely constant (after adjusting for inflation). For further data, see United States Government Accountability Office (2014) on the number of federal civil servants; see United States Department of Defense Manpower Data Center (2015) on the number of active duty military personnel; and see United States Postal Service (2015) on the number of postal workers.
2 This is a quote from Mitch Daniels, former head of the United States Office of Management and Budget.
3 See also Taussig-Rubbo (2009: 121) on the bipartisan support for this "reinventing" of government.
4 Upon taking office, President Obama called for an overhaul of federal contracting in an effort to cut $40 billion from total annual expenditures of $500 billion (Zeleny 2009). By 2014, the total annual expenditure had been reduced to $460 billion (Ivory 2014).
5 The reforms enacted by the Obama administration—to "maximize the use of full and open competition" (Obama 2009)—are seemingly premised on the notion that government contracting is failing to meet some of the goals of his predecessors, *not* that the goals themselves are misguided.
6 The contractor-to-soldier ratios on the battlefield across US wars for which statistics are available are: the Revolutionary War, 1:6; the Mexican-American War, 1:6; the Civil War, 1:5; WWI, 1:24; WWII, 1:7; Korea, 1:2.5; Vietnam, 1:5; the Gulf War, 1:25; Bosnia, 1:1. See Frisk and Trunkey (2008: 13).
7 Blackwater changed its name to Xe Services in 2009 amid controversy and then to Academi in 2011. To avoid confusion, I will continue to refer to this private military firm as Blackwater throughout this book.
8 For similar language on "the sound of battle," see Blizzard (2004: 8–9).
9 See Percy (2006: 14), who problematizes the distinction between "offensive" and "defensive" combat.
10 For an account of the planned coup, see Roberts (2006).
11 Mann was later extradited to Equatorial Guinea and, in July of 2008, sentenced to 34 years in prison.

12 By the "provision of goods," I mean to include all welfare-enhancing provisions, be they the provision of material resources or services. Occasionally, I will examine services alone, leaving materials to the side. When I do so, however, I clearly flag that my analysis is concerned with services and not the broader category of "goods."

13 Wulf (2008: 192) draws a similar distinction between authorization and supply, referring to the former as "bottom-up privatization" and the latter as "top-down privatization." See also Fabre (2012: 187) who distinguishes between "authorization" and "prosecution" and Pattison (2014: 206) who distinguishes between "authorization" and "provision."

14 For this point, see Keegan (1994: 5) and Avant (2005: 3).

15 For this list of excluded wars bar one, see Coady (2008: 4). Coady calls the Rwandan genocide a tribal war. The difficulty is that Hutus and Tutsis were ethnic groups, not tribes. For a lucid critique on how the sort of mistake that Coady makes frustrated efforts of foreign assistance during the genocide, see Power (2013: 355–356). To avoid this mistake, I supplement Coady's list with the further category of genocidal war, under which I include the Rwandan genocide.

16 For an alternative approach, see Reitberger (2013: 69), who follows a definition offered in the Correlates of War Project, requiring "1000 battle-related fatalities per year" to count as a war. See Parry (2015: 185 fn32) for skepticism about this approach.

17 For a further discussion of this concept, see Coady (2008: 5–7).

18 Advocates of the "conventional" approach to the ethics of war are often said to endorse a bifurcated morality while advocates of the "revisionist" approach are said to endorse a more unified morality. For the conventional approach, see Walzer (2006b, 2006c), Benbaji (2008), and Emerton and Handfield (2009). For the revisionist approach, see most notably Rodin (2002), McMahan (2004, 2006a, 2006b, 2008, 2009), Coady (2008a), and Fabre (2012). The edited volume by Rodin and Shue (2008) comprehensively sketches the contours of this debate. Lazar (2012) offers a sharp critical review.

19 Unless otherwise specified, I use the phrases "military action," "use of force," "provision of military force," and "provision of force" as synonymous with war.

20 See Dorfman and Harel (forthcoming: 1–2) for similar language.

21 The lease of Hessian soldiers, as they came to be known, was so profitable that it funded, almost entirely, the Hesse-Kassel annual budget, serving as their primary source of internal revenue. It also subsidized Frederick's extravagant lifestyle. See Singer (2007: 33).

22 More details about the various stages of privatization will be sketched in Section 1.6.

23 This is a methodological point not to be confused with substantive contextualism (Walzer 1983; Miller 2002). I am not suggesting that principles of justice deviate from sphere to sphere. Rather I am only suggesting, more modestly, that the abstract determination of which goods a government must guarantee its citizens and which must be provided by the government itself will be less effective, as a piece of political theory, than a careful examination of one good, in this case military force, because of the complex minutiae involved.

24 I follow Pattison (2014) and others who refer to both as "private military and security companies" (PMSCs). Distinguishing the supposedly non-military operations of PSCs from the military operations of PMCs is confusing and for my purposes unnecessary.

25 See Coates (1997: 123–145), Orend (2000: 192–194; 2006: 50–51), Caney (2005: 205–206), Held (2005), Thompson (2005), McPherson (2007), Fabre (2010; 2012: 112–118, 141–156; 2014: 401–409), Pattison (2008a, 2014: 205–220), Finlay (2010, 2013), Schwenkenbecher (2012: 86–90; 2013), Reitberger (2013), Parry (2015), and Lazar (forthcoming).

26 See Walzer (2006b: 25–29; 2008), Coady (1992; 2008: 205–227), Sandel (1998; 2009: 75–102), Lynch and Walsh (2000), Pattison (2008, 2010, 2012, 2014), Fabre

(2010, 2012: 208–238), Kasher (2008), Baker (2008, 2011a, 2011b), Harel (2008; 2011; 2014: 65–106), Dorfman and Harel (2013), and Steinhoff (2008, 2011).

27 It should be noted that, colloquially, we sometimes say that a private company has outsourced its business to other private companies by hiring them for the provision of certain goods ("Delta has outsourced its call center operation to Sykes Enterprise"). We also say, more loosely, that a private company has outsourced its business to another country when a part of the business is relocated there ("Delta has outsourced its call center operation to India"). In this book, neither usage will be employed. Similarly, I leave aside questions to do with public agencies outsourcing work to other public agencies. My interest in outsourcing is in the interplay between public and private and whether this interplay is morally troubling. With this in mind, the fourfold classification of military force should be clear: either the provision of military force will be purely private, privately outsourced, publicly outsourced, or purely public.

28 Note, however, that these soldiers may not have engaged in action on behalf of BP that would be classified as war. However, to the extent that they were so engaged, this example should suffice.

29 The pure private and pure public variants will henceforth be referred to simply as the private variant and the public variant.

30 Outside of military provision, we see different combinations of these variants with a number of goods. The provision of mail delivery in the US, for example, is in part private (private companies like Microsoft hire FedEx to deliver their mail), in part privately outsourced (Microsoft also hires the US Postal Service to deliver its mail), in part publicly outsourced (some government agencies hire FedEx to deliver their mail) and in part public (some government agencies hire the US Postal Service). Similarly, the provision of primary education in the US is undertaken publicly, privately, and through public outsourcing. Private schools exist alongside public schools, and both exist alongside the burgeoning array of charter schools, which are publicly established but privately run. Finally, to take one last example, health care in the UK is provided both publicly and privately; the state-operated National Health Service (NHS) provides all citizens with health care, but some citizens choose to receive supplemental coverage from private firms, whom they pay for service.

31 Consequently, for any good g, there are 16 possible paths that a society may chart in providing g. The 16 potential scenarios are simply what would result whenever an entity is presented with four discreet items and allowed to select up to four of these items (including zero and four).

32 At any given time in medieval Ireland, there were between 80–100 tuaths. On tuaths, see Binchy (1970), Rothbard (1973: 220), and Peden (1977).

33 Perhaps the best analysis of how the norm against private authorization and private supply developed can be found in Percy (2007a).

34 Above, I noted that privatization refers to the shift from public provision towards private provision. The rendering that I give here is consistent with that rendering. Here, I simply further clarify the shift.

35 As we can see, privatization may involve a shift towards increased private authorization *or* towards increased private supply. But imagine two societies: Society 1 has more private authorization than Society 2, and Society 2 has more private supply than Society 1. Which society, we might ask, has further privatized its military forces? According to my definition, Society 1 has further privatized its military forces. Private authorization is taken to be a weightier indicator of privatization than private supply. This is why I insist that, within a given arrangement $(A_1–A_5)$, privatization refers to "the trend away from public provision of military force towards privately outsourced, publicly outsourced, and ultimately private provision."

36 The non-provision of military force in A_0 is left to the side as it counts neither as privatization nor as nationalization. The considerations that were raised in the previous

footnote about privatization also apply to nationalization (though, of course, nationalization refers to the opposite trend).

37 Given this account, it would be a mistake to think that privatization might refer to strategies that, without increasing the share of privately provided goods to publicly provided goods, make the public provision of goods function *more like* the private provision of goods. Such strategies instead constitute what we may call *marketization*. The use of school vouchers is a quintessential example: by issuing vouchers for education and compelling public schools to compete with one another for these vouchers (or indeed with private schools if the vouchers may be redeemed there), the government can dispose its own schools to behave more like private schools. But, in issuing these vouchers, the government does not, in some unavoidable way, cede services to the private sector. Particularly if we restrict the redemption of vouchers to public schools alone, we might wholly endorse school vouchers without supporting a decrease in the number of schools under government operation or any other diminishment in the government's role as the educator of its citizens—i.e., we need not simultaneously defend privatization while endorsing marketization. This is not to deny that there are critical links between privatization and marketization. As Thomas Frank (2008: 8) describes, markets have been "boring ever deeper into the tissues of the state." Privatization and marketization both represent avenues by which the state, as traditionally understood, has atrophied. Undoubtedly, detailed philosophical research is needed into the whole of this atrophying process. My book, however, will focus exclusively on privatization, leaving these further questions aside.

38 The compensation argument builds upon a position that Nozick (1974: 110–113) develops.

39 Powell was originally quoted in Woodward (1992: 154).

40 What it means to exercise command will be elaborated in Chapter 6.

References

Avant, Deborah (2005). *The Market for Force: The Consequences of Privatizing Security* (Cambridge: Cambridge University Press).

Baker, Deane-Peter (2008). "Of 'Mercenaries' and Prostitutes: Can Private Warriors Be Ethical," in Andrew Alexandra, Deane-Peter Baker, and Marina Caparini (eds), *Private Military and Security Companies: Ethics, Policies, and Civil–Military Relations* (New York: Routledge), pp. 30–42.

Baker, Deane-Peter (2011a). *Just Warriors, Inc.: The Ethics of Privatized Force* (London: Continuum International Publishing).

Baker, Deane-Peter (2011b). "To Whom Does a Private Military Commander Owe Allegiance?," in Paolo Tripodi and Jessica Wolfendale (eds), *New Wars and New Soldiers: Military Ethics in the Contemporary World* (Surrey: Ashgate), pp. 181–198.

Benbaji, Yitzhak (2008). "A Defense of the Traditional Just War Convention," *Ethics*, Vol. 118, No. 3, pp. 464–495.

Binchy, Daniel (1970). *Celtic and Anglo-Saxon Kingship* (Oxford: Oxford University Press).

Bland, Douglas (1999). "A Unified Theory of Civil–Military Relations," *Armed Forced & Society*, Vol. 26, No. 1, pp. 7–25.

Blizzard, Stephen M. (2004). "Increasing Reliance on Contractors on the Battlefield: How Do We Keep from Crossing the Line?," *Air Force Journal of Logistics*, Vol. 28, No. 1, pp. 3–13.

Caney, Simon (2005). *Justice Beyond Borders: A Global Political Theory* (Oxford: Oxford University Press).

Clausewitz, Carl von (2008 [1832]). *On War*, translated and edited by Michael Eliot Howard and Peter Paret (Oxford: Oxford University Press).

Coady, C.A.J. (1992). "Mercenary Morality," in A.G.D. Bradney (ed), *International Law and Armed Conflict* (Stuttgart: Franz Steiner Verlag), pp. 55–70.

Coady, C.A.J. (2008). *Morality and Political Violence* (Cambridge: Cambridge University Press).

Coates, A.J. (1997). *The Ethics of War* (Manchester: Manchester University Press).

DeYoung, Karen (2007). "Other Killings by Blackwater Staff Detailed," *Washington Post*, 2 October.

Dorfman, Avihay and Alon Harel (2013). "The Case Against Privatization," *Philosophy and Public Affairs*, Vol. 41, No. 1, pp. 67–102.

Dorfman, Avihay and Alon Harel (forthcoming). "Against Privatisation As Such," *Oxford Journal of Legal Studies*, pp. 1–28.

Emerton, Patrick and Toby Handfield (2009). "Order and Affray: Defensive Privileges in Warfare," *Philosophy and Public Affairs*, Vol. 37, No. 4, pp. 382–414.

Fabre, Cécile (2010). "In Defence of Mercenarism," *British Journal of Political Science*, Vol. 40, No. 3, pp. 539–559.

Fabre, Cécile (2012). *Cosmopolitan War* (Oxford: Oxford University Press).

Fabre, Cécile (2014). "Rights, Justice, and War: A Reply," *Legal Theory*, Vol. 33, No. 3, pp. 391–425.

Fainaru, Steve (2007). "Iraq Contractors Face Growing Parallel War," *Washington Post*, 16 June.

Finlay, Christopher J. (2010). "Legitimacy and Non-State Political Violence," *The Journal of Political Philosophy*, Vol. 18, No. 3, pp. 287–312.

Finlay, Christopher J. (2013). "Fairness and Liability in the Just War: Combatants, Non-Combatants, and Lawful Irregulars," *Political Studies*, Vol. 61, No. 1, pp. 142–160.

Frank, Thomas (2008). *The Wrecking Crew: How Conservatives Rule* (New York: Metropolitan Books).

Frisk, Daniel and R. Derek Trunkey (2008). *Contractors Support of US Operations in Iraq*. Report for the Senate Committee on the Budget, Publication Number 3053 (Washington, DC: Congressional Budget Office).

Glanz, James (2008). "Report on Iraq Security Lists 310 Contractors," *New York Times*, October 28.

Harel, Alon (2008). "Why Only the State May Inflict Criminal Sanctions: The Case Against Privately Inflicted Sanctions," *Legal Theory*, Vol. 14, No. 2, pp. 113–133.

Harel, Alon (2011). "Outsourcing Violence?," *Law and Ethics of Human Rights*, Vol. 5, No. 2, pp. 396–413.

Harel, Alon (2014). *Why Law Matters* (Oxford: Oxford University Press).

Held, Virginia (2005). "Legitimate Authority in Non-State Groups Using Violence," *Journal of Social Philosophy*, Vol. 36, No. 2, pp. 175–193.

Isenberg, David (2009). *Shadow Force: Private Security Contractors in Iraq* (Westport: Praeger Security International).

Ivory, Danielle (2014). "Federal Contracts Plunge, Squeezing Private Companies," *The New York Times*, January 15.

Kasher, Asa (2008). "Interface Ethics: Military Forces and Private Military Companies," in Andrew Alexandra, Deane-Peter Baker, and Marina Caparini (eds), *Private Military and Security Companies: Ethics, Policies, and Civil–Military Relations* (New York: Routledge), pp. 235–246.

Keegan, John (1994). *A History of Warfare* (New York: Vintage Books).

Lazar, Seth (2012). "The Morality and Law of War," in Andrei Marmor (ed), *The Routledge Companion to Philosophy of Law* (New York: Routledge), pp. 364–379.

Lazar, Seth (forthcoming). "Authorisation and the Morality of War," *Australasian Journal of Philosophy*, pp. 1–19.

Light, Paul (2006). "The New True Size of Government." Available at https://wagner.nyu.edu/files/performance/True%20Size%20Research%20Brief.pdf (accessed December 28, 2015).

Lynch, Tony and A.J. Walsh (2000). "The Good Mercenary?," *The Journal of Political Philosophy*, Vol. 8, No. 2, pp. 133–153.

McMahan, Jeff (2004). "The Ethics of Killing in War," *Ethics*, Vol. 114, No. 4, pp. 693–733.

McMahan, Jeff (2006a). "On the Moral Equality of Combatants," *The Journal of Political Philosophy*, Vol. 14, No. 4, pp. 377–393.

McMahan, Jeff (2006b). "The Ethics of Killing in War," *Philosophia*, Vol. 34, No. 1, pp. 23–41.

McMahan, Jeff (2008). "The Morality of War and the Law of War," in David Rodin and Henry Shue (eds), *Just and Unjust Warriors: The Legal and Moral Status of Soldiers* (Oxford: Oxford University Press), pp. 19–43.

McMahan, Jeff (2009). *Killing in War* (Oxford: Oxford University Press).

McPherson, Lionel (2007). "Is Terrorism Distinctively Wrong?," *Ethics*, Vol. 117, No. 3, pp. 524–546.

Miller, David (2002). "Two Ways to Think About Justice," *Philosophy, Politics, and Economics*, Vol. 1, No. 1, pp. 5–28.

Moorhead, Molly (2012). "Contract Spending on the Decline," *Tampa Bay Times Politifact*, December 12. Available at www.politifact.com/truth-o-meter/promises/obameter/promise/432/cut-federal-contracts/ (accessed December 28, 2015).

More, Thomas (2002 [1516]). *Utopia*, edited by George M. Logan and Robert M. Adams (Cambridge: Cambridge University Press).

Nozick, Robert (1974). *Anarchy, State, and Utopia* (New York: Basic Books).

Obama, Barak (2009). "Memorandum for the Heads of Executive Departments and Agencies: Subject: Government Contracting." Available at www.whitehouse.gov/21stcentury gov/actions/reforming-government-contracting (accessed December 28, 2015).

Orend, Brian (2000). *War and International Justice: A Kantian Perspective* (Waterloo: Wilfrid Laurier University Press).

Orend, Brian (2006). *The Morality of War* (Toronto: Broadview Press).

Pan, Ester (2004). "Iraq: Military Outsourcing," *Backgrounder for the Council on Foreign Relations*. Available at www.cfr.org/publication/7667/ (accessed December 28, 2015).

Parry, Jonathan (2015). "Just War Theory, Legitimate Authority, and Irregular Belligerency," *Philosophia*, Vol. 43, No. 1, pp. 175–196.

Pattison, James (2008a). "Just War Theory and the Privatization of Military Force," *Ethics and International Affairs*, Vol. 22, No. 2, pp. 143–162.

Pattison, James (2010a). "Deeper Objections to the Privatisation of Military Force," *The Journal of Political Philosophy*, Vol. 18, No. 4, pp. 425–447.

Pattison, James (2012). "The Legitimacy of the Military, Private Military and Security Companies, and Just War Theory," *European Journal of Political Theory*, Vol. 11, No. 2, pp. 131–154.

Pattison, James (2014). *The Morality of Private War* (Oxford: Oxford University Press).

Peden, Joseph R. (1977). "Property Rights in Celtic Irish Law," *Journal of Libertarian Studies*, Vol. 1, No. 2, pp. 81–95.

Percy, Sarah (2006). *Regulating the Private Security Industry*, Adelphi Paper (New York: Routledge).

Percy, Sarah (2007a). *Mercenaries: The History of a Norm in International Relations* (Oxford: Oxford University Press).

Percy, Sarah (2007b). "Morality and Regulation," in Simon Chesterman and Chia Lehnardt (eds), *From Mercenaries to Market: The Rise and Regulation of Private Military Companies* (Oxford: Oxford University Press), pp. 11–28.

Polgreen, Lydia (2008a). "Fact Mirrors Fiction in African Coup Trial," *New York Times*, June 18.

Polgreen, Lydia (2008b). "British Mercenary Sentenced in Coup Plot," *New York Times*, July 8.

Power, Samantha (2013). *"A Problem from Hell": America and the Age of Genocide*, second edition (New York: Basic Books).

Reitberger, Magnus (2013). "License to Kill: Is Legitimate Authority a Requirement for Just War?," *International Theory*, Vol. 5, No. 1, pp. 64–93.

Roberts, Adam (2006). *The Wonga Coup: Guns, Thugs, and a Ruthless Determination to Create Mayhem in an Oil-Rich Corner in Africa* (London: Profile Books).

Rodin, David (2002). *War and Self-Defense* (Oxford: Oxford University Press).

Rodin, David and Henry Shue (eds) (2008). *Just and Unjust Warriors: The Legal and Moral Status of Soldiers* (Oxford: Oxford University Press).

Rothbard, Murray N. (1973). *For a New Liberty: The Libertarian Manifesto* (New York: Macmillan).

Runzo, Joseph (2008). "Benevolence, Honourable Soldiers, and Private Military Companies: Reformulating Just War Theory," in Andrew Alexandra, Deane-Peter Baker, and Marina Caparini (eds), *Private Military and Security Companies: Ethics, Policies, and Civil–Military Relations* (New York: Routledge), pp. 56–69.

Sandel, Michael (1998). "Commodification, Commercialization, and Privatization," *What Money Can't Buy: The Moral Limits of Markets*, The Tanner Lectures on Human Value. Paper presented at Brasenose College, Oxford, May 11–12.

Sandel, Michael (2009). *Justice: What's the Right Thing to Do?* (New York: Farrar, Strauss, and Giroux).

Schemo, Diana Jean (1996). "Oil Companies Buy an Army to Ward Off Rebels in Colombia," *New York Times*, August 22.

Schwartz, Moshe and Joyprada Swain (2011). *Department of Defense Contractors in Afghanistan and Iraq: Background and Analysis*. Report for Congress, R40764, May 13 (Washington, DC: Congressional Research Service). Available at http://fas.org/sgp/crs/natsec/R40764.pdf (accessed December 28, 2015).

Schwartz, Moshe and Wendy Ginsberg (2013). *Department of Defense Trends in Overseas Contract Obligations*, Report for Congress, R41820, March 1 (Washington, DC: Congressional Research Service). Available at http://fas.org/sgp/crs/natsec/R41820.pdf (accessed December 28, 2015).

Schwellenbach, Nick (2014). "Is the Federal Civilian Workforce Really Growing? Some Important Context." Available at: www.foreffectivegov.org/is-federal-civilian-workforce-really-growing-some-important-context (accessed December 28, 2015).

Schwenkenbecher, Anne (2012). *Terrorism: A Philosophical Inquiry* (Basingstoke: Palgrave Macmillan).

Schwenkenbecher, Anne (2013). "Rethinking Legitimate Authority," in Fritz Allhoff, Nicholas G. Evans, and Adam Henschke (eds), *Routledge Handbook of Ethics and War: Just War Theory in the Twenty-First Century* (Oxford: Routledge), pp. 161–170.

Shorrock, Tim (2015). "The Afghanistan War is Still Raging—but This Time It's Being Waged by Contractors," *The Nation,* February 4. Available at www.thenation.com/blog/197057/afghanistan-war-still-raging-time-its-being-waged-contractors (accessed December 28, 2015).

Singer, Peter Warren (2005). "Outsourcing War," *Foreign Affairs,* Vol. 84, No. 2, pp. 119–132.

Singer, Peter Warren (2007). *Corporate Warriors: The Rise of the Privatized Military Industry,* updated edition (Ithaca: Cornell University Press).

Steinhoff, Uwe (2008). "What Are Mercenaries?," in Andrew Alexandra, Deane-Peter Baker, and Marina Caparini (eds), *Private Military and Security Companies: Ethics, Policies, and Civil–Military Relations* (New York: Routledge), pp. 19–29.

Steinhoff, Uwe (2011). "Ethics and Mercenaries," in Paolo Tripodi and Jessica Wolfendale (eds), *New Wars and New Soldiers: Military Ethics in the Contemporary World* (Surrey: Ashgate), pp. 137–151.

Taussig-Rubbo, Mateo (2009). "Outsourcing Sacrifice: The Labor of Private Military Contractors," *Yale Journal of Law & the Humanities,* Vol. 21, No. 1, pp. 101–164.

Thompson, Janna (2005). "Terrorism, Morality, and Right Authority," in Georg Meggle (ed), *Ethics of Terrorism & Counter-Terrorism* (Frankfurt: Ontos Verlag), pp. 151–160.

United States Department of Defense Manpower Data Center (2015). "DoD Personnel, Workforce Reports & Publications." Available at www.dmdc.osd.mil/appj/dwp/dwp_reports.jsp (accessed December 28, 2015).

United States Government Accountability Office (2014). "Federal Workforce: Recent Trends in Federal Civilian Employment and Compensation." Available at www.gao.gov/assets/670/660449.pdf (accessed December 28, 2015).

United States House Committee on Government Reform (2006). *Dollars Not Sense: Government Contracting Under the Bush Administration* (Washington, DC: Government Printing Office).

United States Postal Service (2015). "Number of Postal Employees Since 1926." Available at https://about.usps.com/who-we-are/postal-history/employees-since-1926.pdf (accessed December 28, 2015).

Walzer, Michael (1983). *Spheres of Justice: A Defense of Pluralism and Equality* (New York: Basic Books).

Walzer, Michael (2006a). "Regime Change and Just War," *Dissent,* Vol. 53, No. 3, pp. 103–108.

Walzer, Michael (2006b). *Just and Unjust Wars: A Moral Argument with Historical Illustrations,* 4th edition (New York: Basic Books).

Walzer, Michael (2006c). "Response to McMahan's Paper," *Philosophia,* Vol. 34, No. 1, pp. 43–45.

Walzer, Michael (2008). "Mercenary Impulse: Is there an Ethic that Justifies Blackwater?," *The New Republic,* March 12, pp. 20–21.

Woodward, Bob (1992). *The Commanders* (New York: Simon and Schuster).

Wulf, Herbert (2008). "Privatization of Security, International Interventions, and the Democratic Control of Armed Forces," in Andrew Alexandra, Deane-Peter Baker, and Marina Caparini (eds), *Private Military and Security Companies: Ethics, Policies, and Civil–Military Relations,* (New York: Routledge), pp. 191–202.

Zeleny, Jeff (2009). "Obama to Change Contract Awarding," *New York Times,* March 4.

Part I

Authorizing war

2 Legitimate authority and monopolization

Ought governments to be the ultimate arbiters in decisions about war or should some of these decisions be left to private actors? Just war theorists, beginning with Augustine and Aquinas, have conventionally defended a requirement limiting the authorization of war to political sovereigns.[1] This restriction, along with the *jus ad bellum* requirements of just cause and right intention, were considered necessary conditions for the justified authorization of war (Johnson 1999: 42).[2] While the requirement of legitimate authority has long been central to just war theory, contemporary scholars have tended to afford it less scrutiny than other *jus ad bellum* requirements.[3] The result is under-theorization of the requirement's moral grounds and insufficient elaboration of the properties that must be possessed, if any, by entities to satisfy its demands.[4]

This chapter begins by elaborating a conception of legitimate authority, which empties the concept of linguistic and substantive baggage that has accrued over centuries in the just war tradition.[5] The conception that I offer does *not* assume that only entities with statehood (or political institutions) possess the right to authorize military force. Such an assumption would foreclose inquiry into the *private* possession of legitimate authority and, without defense, would be unjustifiable (Caney 2005: 205; Held 2005: 185; Thompson 2005: 91; McPherson 2007: 542; Reitberger 2013: 67; Schwenkenbecher 2013: 163; Parry 2015: 184–185). Instead, I offer a more barren conception, which defines legitimate authority as a liberty-right and claim-right to authorize military force. After elaborating this conception in Section 2.1, I outline the concept of "monopolization over military authorization" in Section 2.2.[6] Proponents of private military authorization reject public monopolization, while opponents defend it. Sections 2.3, 2.4, and 2.5 consider and reject a prominent argument *against* public monopolization: Cécile Fabre's (2008; 2012: 112–118, 141–156; 2014: 401–409) cosmopolitan critique.[7] Consideration of this influential argument will set the stage for a positive defense of public monopolization in Chapters 3 and 4.

2.1 Legitimate authority

Let us begin then with the concept of legitimate authority. One set of wartime worries that we must resolve are what Arthur Isak Applbaum (2007: 360–361) calls our "first-order moral considerations": to plumb these before going to war, we ask (as the entity considering war) "how much death, destruction, and misery will be inflicted on their soldiers and ours, for what reasons and for whose benefit, and for how long and at what cost and with what prospects of success?." A second set of considerations mulls not over the substance of arguments for war but over who may assess and act upon the substance of these arguments—the entity, in other words, "who is to *decide* upon the first-order moral judgments" and authorize war. It is this second set of considerations that concerns us.[8]

Entities that are justified in deciding and acting upon the first-order moral judgments of war will be said to possess *legitimate authority* to authorize military force. An entity that possesses legitimate authority, on my account, has both a liberty-right to authorize military force *and* a claim-right that those who lack such a liberty-right refrain from forcefully stymieing their efforts.[9] A liberty-right to perform some action *x* entails that the right-bearer has no duty not to perform *x*; by contrast, a claim-right to perform *x* entails that others are under some sort of duty with respect to the performance of *x* (Hohfeld 1919: 36–45). The duty under which one is placed by a claim-right may be a duty to perform some action or a duty to abstain in the performance of some action—as we can see, the duty that is generated by the claim-right to authorize military force is of the latter variety (Wenar 2015: 6).[10] Thus, an entity with the legitimate authority to authorize war has no duty *not* to authorize war (a liberty-right), and such an entity is owed a duty of forbearance by those who lack such a liberty-right to refrain from forcefully obstructing their war efforts (a claim-right).

Of course, with some restrictions flowing from our further just war commitments, other groups with legitimate authority may indeed seek to forcefully obstruct their war efforts. After all, a war just is the forceful obstruction of the enemy's war efforts. But those who lack a liberty-right to authorize war are precluded from forcefully obstructing the war efforts of those who possess the liberty-right (and/or claim-right) to authorize war.

My account remains agnostic on whether legitimate authority ought to function as a stand-alone just war criterion or as a component of other *jus ad bellum* criteria. For some, legitimate authority is thought to perform "extra work" so that even entities meeting other criteria might still lack justification to authorize military force (Finlay 2010: 301). For others, legitimate authority is considered constitutive of one or more further criteria—whether just cause, right intention, likelihood of success, proportionality, or last resort. Without always defending a conception of legitimate authority that takes it to be constitutive of other just war criteria, several theorists have pointed to the way in which legitimate authority at least *bears on* the fulfillment of other requirements. Suppose for the sake of argument that legitimate authority demands democratic consultation. We might say that, in a war of political independence, claims to just cause "would be falsified if the community in question preferred to remain part of a larger political entity" (Finlay 2010: 305).[11] Right intention may be doubtful insofar as the failure to consult (and perhaps unwillingness to consult) indicates alternative (and unjustifiable) aims. Likelihood of success would be diminished in the absence of popular support (Fabre 2008: 971; Reitberger 2013: 91; Schwenkenbecher 2013: 165; Gross 2013: 222). The negative consequences of bypassing political processes could threaten the proportionality of war (Finlay 2010: 307; Reitberger 2013: 91; Lazar forthcoming). And failure of consultation may contravene the principle of last resort since consultation is, after all, one measure that a group could take before resorting to war (Finlay 2010: 208; Fabre 2012: 150; Reitberger 2013: 81). But regardless of how legitimate authority bears on these other just war requirements, a successful defense of legitimate authority should maintain its force—whether the requirement is capable of standing on its own or collapses into a second just war criterion.

It should finally be noted that my account treats legitimate authority as a deontological requirement of war. The authorization of military force in the absence of legitimate authority is wrong, not because it generates bad outcomes, but because it violates the rights of others.[12] Potential authorizing entities have a duty to possess (or acquire) legitimate authority even if the *consequences* of authorizing war without legitimate authority are superior to alternatives. As with any deontological constraint—in war (e.g., not purposefully killing civilians or torturing soldiers) and outside war (e.g., not assaulting innocent people or stealing their belongings)—difficult questions arise about whether individuals may *ever* permissibly violate the rights of others. Can we firebomb a city to save our civilization (in what Walzer calls "supreme emergencies") or torture a terrorist to prevent nuclear attack?[13]

While any fleshed out theory of legitimate authority must consider these circumstances, we should remember that rights-based deontological constraints are not our only moral commitments. They are not "the whole of morality," as David Rodin (2002: 28) argues, and "ascribing a right is importantly different to asserting an all-things-considered moral judgment." To ascribe a right is to assign special value to the interest or will that it protects—it is to insist that such interest or will "trumps" goal-directed consequentialist considerations (Dworkin 1984).[14] As a result, when rights are violated, moral loss accrues. For instance, even if one is justified in (or excused from) burning a neighbor's field to save a village, the "neighbor's claim-right against having his property destroyed does not disappear" and "the neighbor is still owed some compensation, redress, or apology" (Rodin 2002: 28). The requirement of legitimate authority, on my conception, functions like other deontological constraints. It need not constitute an absolute ("until the heavens fall") demand. But it must be resistant to goal-directed exchanges of rights-violations for maximization of the greater good. The challenge is to identify *why* military authorization is rights-violating in the absence of legitimate authority.

2.2 Monopolization

While an account of legitimate authority is vital for a theory of military privatization, it is not sufficient. As alluded to in Section 1.7, an entity may possess legitimate authority to authorize war but *still* lack justification to use force short-of-war to prevent individuals or groups who lack legitimate authority from authorizing *their own* wars.[15] If an entity is to take such preventive measures against those who lack legitimate authority, it must possess, at a minimum, the liberty-right to do so. A stronger justification would be that the entity possesses a claim-right that those who lack legitimate authority *refrain* from stymieing their preventive efforts. And an even stronger justification would be that, within a specified jurisdiction, *no one* other than (perhaps) the entity in question *in fact* possesses legitimate authority, and *all* are subject to the aforementioned liberty- and claim-rights possessed by that entity. For the purposes of my analysis, the *authority to monopolize the authorization of military force* within a jurisdiction will map onto this third understanding: as a conjunction of the liberty-right to prevent all others within that jurisdiction from authorizing war and the claim-right that all others within that jurisdiction refrain from stymieing their preventive efforts.[16] As should now be clear, an entity may possess the authority to authorize military force but lack the authority to monopolize military authorization or lack the authority to authorize military force but possess the authority to monopolize military authorization (using force short-of-war), or both, or neither.

With this account in mind of what it means for an entity to possess the authority to monopolize the authorization of military force within a jurisdiction, let me briefly make two points of clarification. First, this right of monopolization is neither entailed by nor does it entail the right to monopolize the authorization of *all* force. A group, for example, might lack what John Locke (2002 [1690]: ch. 12) calls the *federative right* to monopolize the authorization of force—this is an exclusive right to prevent the private authorization of force against *non-group members*—but possess the exclusive right to prevent the private authorization of force against group members. Just as the governing board of a neighborhood association might be entitled to issue sanctions *only* against those who choose to build their homes in that neighborhood, so too might a group be entitled to prevent force by members of that group only against other members of *that* group. And attempts by the group to prevent members from authorizing force against non-members may be no less out of place than attempts by a neighborhood association to keep residents from interacting in certain ways with non-residents. Both organizations, though it may sound otherworldly today, may be charged solely with governing relations between members, remaining impotent in the governance of all other relations. Many, of course, following in the tradition of Max Weber (2004 [1919]: 33), have defined the state by its capacity to monopolize the use of *all* force, thus encouraging an analysis of federative power alongside, and often as indistinguishable from, non-federative power. But we must not allow this common conflation to blind us from the very simple point that federative and non-federative rights are not necessarily possessed or lacked together.

Perhaps less obviously, we may ask—even of groups who lack the federative right to monopolize the authorization of force—whether they may monopolize the authorization of all *forms* of force between group members or only of some forms of force. It may be that a group is permitted, for instance, to monopolize the authorization of punishment for homicides committed by and against group members but not to monopolize the authorization of war by and against group members.[17] A position that endorses a monopoly on the use of non-military force but not military force may be normatively indefensible, but the conceptual possibility should be acknowledged.[18]

2.3 Cosmopolitan critique of public military monopolization

Having now clarified the concepts of legitimate authority and monopolization, I will turn to Cécile Fabre's cosmopolitan challenge of public monopolization over military authorization.[19] Her challenge is perhaps the most influential critique in contemporary political theory, and so I will examine it in some detail before turning to my positive case for military monopolization. There are at least two discernible versions of her critique. A *stronger version* is defended in earlier work (Fabre 2008), and a *weaker version* is defended in later work (Fabre 2012: 142–156; 2014: 401–409). Both claim to jettison the principle of legitimate authority and do so according to a common central argument, but they arrive at somewhat different conclusions. The common central argument from which each version of the critique begins is this:

> [T]he right to protect oneself from violations of one's human rights on the part of others is also a *human* right, in the sense that it is a right to a freedom (the freedom to defend oneself without interference, against others), which we need in order to lead a minimally decent life. By extension, the right to wage a war in defense of one's human rights should also be conceived of as a human right.
>
> (Fabre 2012: 145)[20]

Fabre's argument, as we can see, contains two steps. The first step, which may be referred to as the *individual-claim*, is that individuals have rights to authorize deadly force under certain conditions (for example, in self-defense). The second step, which may be referred to as the *military-claim*, is that the grounds upon which individuals are entitled to authorize deadly force apply (in just this way) to private groups of individuals or single individuals authorizing deadly force on the scale that wars are fought.[21]

What distinguishes the stronger challenge from the weaker challenge is the emphasis on consent. The stronger challenge does not require that authorizing entities seek the consent of those on behalf of whom war is authorized—call these individuals the *constituents* of the authorizing entity. The permission for private military authorization applies *regardless* of whether constituents consent to authorization, or indeed whether they were granted the opportunity to decide in the first place.[22]

The weaker challenge, by contrast, requires that authorizing entities, at a minimum, seek consent. The permission for private military authorization applies *only if* consent cannot be obtained. Moreover, according to the weaker challenge, in those instances when consent cannot be obtained, private authorizing entities must consider whether constituents would have consented. And they must make themselves available following military authorization for review and possible punishment by constituents to determine whether the presumption of consent was correct or erroneous. Fabre (2012: 155) specifies these requirements as follows:

> When conditions are not there under which victims of rights violations can give consent (indirect or not) either to the war itself or to institutions mandated with fighting the war, insurgents may take matters into their own hands, subject to two conditions: they have good reason to believe that their fellow community members *would* consent if they could, *and* they put into place institutional mechanisms whereby those for whose sake they fight can hold them accountable once the war is over.

Private military authorization, according to the weaker challenge, thus serves as a back-up permission. Only when consent cannot be obtained is this backup-permission triggered and the central argument (comprised of the individual- and military-claims) underpinning this permission applicable.

2.4 Stronger cosmopolitan critique

Let us consider the merits of each critique. In keeping with the stronger cosmopolitan critique, few would deny the individual-claim. Though the list of permissible instances of private force are debated—whether, for example, the right to punish is attributed to and is exercisable by individuals is by no means obvious—the existence of *some* instances is not. By contrast, the notion that an individual right to authorize deadly force entails a right to authorize *military* force is far less intuitive. At least three objections may be raised against the military-claim. Only the final objection, I will argue, is ultimately decisive, but each should be considered to understand precisely why the final objection is damaging. Insofar as the final objection undermines the military-claim and insofar as that claim is needed to justify the back-up permission of private military authorization in the weaker cosmopolitan critique, the final objection also threatens the weaker critique.[23]

2.4.1 Military preparation

The first objection to the military-claim proceeds with reference to weapons ownership. According to this objection, an individual right to self-defense may justify the ownership of guns. But just as such a right does not necessarily entail the further right to stockpile certain weapons, the individual right to self-defense does not necessarily entail the further right to undertake the *preparations that would be necessary* for military authorization. Hugh LaFollette (2000: 268) persuasively reminds us that the individual right to self-defense may be robust but not sufficiently robust to accommodate the individual right to prepare for confrontations by, say, accumulating tactical nuclear or biochemical weapons. Even if we could imagine extreme scenarios in which the private deployment of a tactical nuclear or biological weapon would be justifiable in self-defense (if that were possible), private citizens should not be permitted to anticipate these scenarios by taking the precaution of obtaining such weapons.[24] Weapons are extremely dangerous, and "[t]okens of fundamental rights may be restricted to protect others from serious harms arising from the exercise of those rights" (LaFollette 2000: 265).

An argument structured precisely along these lines is available to the critic of private military authorization. Private actors require weapons and training if they are to have a reasonable prospect of success when authorizing military force (which, of course, is demanded of just wars). A state that instituted a ban on weapons—on tactical nuclear weapons, biological weapons, perhaps even grenades and guns—would seemingly be a state whose residents were incapable of privately authorizing military force. Similarly, a state that instituted a ban on military training—on the formation of organizations designed to authorize military force, the organization of members into cells or units, and the implementation of disciplinary regimens to prepare members for combat—would also be a state, it seems, whose residents were incapable of privately authorizing military force. If the substance of our right to self-defense may be limited to protect others from harm, as LaFollette correctly argues (I believe), then the tools and preparation upon which private military authorization is dependent, both empirically and morally, may be outlawed. This, in turn, may obviate the need to query private military authorization.

At least two related difficulties, however, confront this position. First, governments may never be capable of outlawing the totality of weapons and training that are needed for war. Second, even if the prohibition on conventional weaponization (assault rifles and training camps) were successfully implemented, enterprising militants may rely upon unconventional strategies. We need only recall the destruction that commercial airliners, transformed into precision guided missiles, caused on September 11 to see how conventional tools, which should not be banned, may be converted into unconventional weapons for the authorization of war. An airplane, we might say, is a *dual-use* machine, which may be used for peaceful purposes and war making purposes.[25] By the same token, the training that is required for war may be dual-use training. A computer software engineer may develop the skills that are needed to initiate cyber-attacks, which take down electrical grids. But, of course, a government ought not to prohibit the education of computer software engineers. Because a morally permissible government prohibition on weapons and training cannot rule out dual-use weapons and training, such a governmental prohibition will be incapable of grounding public monopolies over military authorization.[26]

2.4.2   Emergency authorization

A second potential objection to the military-claim is that arguments for public military monopolization are *institutional*, not *interpersonal* propositions. As LaFollette (2000: 270) writes about gun control, it "does not concern what private individuals should do but what governments should allow private individuals to do." Perhaps we should say the same of public military monopolization—that it concerns what governments should allow private individuals to do, not what private individuals ought to do. While this view, I believe, is largely correct, it camouflages a complexity that, once acknowledged, resuscitates the military-claim.

Consider a quintessential example that is used to justify propositions like the military-claim: the Warsaw Ghetto uprising. As Rodin (2002: 139) remarks, when the Jews trapped in the Warsaw Ghetto resisted Nazi aggression, they "were quite literally fighting for their lives" in a way that it becomes possible "to understand their actions as justified wholly within the conceptual scheme of individual rights."[27] Surely, individuals in such a position are justified in privately authorizing military force. Call this position the *argument for emergency authorization.*

The claim that arguments for public military monopolization are institutional, rather than interpersonal propositions, purportedly skirts this challenge. But while the argument for emergency authorization certainly may press an interpersonal question about whether private individuals are ever permitted to authorize military force, it also presses an institutional question that, in a way, precedes the interpersonal question. The question concerns the mechanism by which private actors, like members of the Warsaw Ghetto, must authorize military force. Should they do so as independent private actors or in some other fashion, perhaps even as a public actor?

I noted in Section 1.4 that, while "public authorization" may be equated with "governmental authorization" until more could be said, non-governmental authorization is not always private—for instance, when guerilla movements authorize military force. I follow Pattison (2014: 205), who takes "public" to connote the collective or group (and "public interests," collective interests), and "private" to connote the individual (and "private interests," individual interests). Military authorization is said to be public when the authorizing entity attempts to reflect collective preferences and properly submits to these preferences. Military authorization is private when the authorizing entity does not attempt to reflect collective preferences and/or fails to properly submit to these preferences. The question of legitimate authority in the Warsaw Ghetto uprising is whether the authorizing entity must possess *some set* of properties to authorize military force—and, more particularly, whether they must possess the property of collective authorization. If the answer to the first question is yes, and the authorizing entity must possess *some set* of (yet-to-be-elaborated) properties, then the requirement of legitimate authority stands. If the answer to the second question is yes, and the authorizing entity must possess some set of (yet-to-be-elaborated) *collective* properties, then the challenge to *private* military authorization stands.

Supposing that there are *some* steps available to residents of the Warsaw Ghetto that could transform a private military authorization into a public military authorization, then residents are confronted with a variety of institutional considerations. Should the insurgency, for instance, monopolize the authorization of military force by Ghetto residents? Or might private individuals in the Ghetto be justified in authorizing their own wars of liberation without consulting other Ghetto residents? Once we recognize that the decision to authorize military force by Ghetto residents may be public or private, we must determine whether the private authorization of war would be justifiable or whether public, but nongovernmental actors, are permitted to withdraw decisions of military authorization from the private sector and reserve them for public discharge. In other words, the military-claim remains alive as an institutional proposition and cannot be sidestepped, as LaFollette does in the context of gun control, merely by shelving interpersonal concerns.[28]

2.4.3 Intuitions underpinning the military-claim

The final, and I believe decisive, retort to the argument for emergency authorization, and thus to the military-claim, is that it is not at all obvious that individuals in emergency circumstances, like the Warsaw Ghetto, should be permitted to authorize military force without attempting to consult with other members (and potentially authorizing military force publicly). The military-claim relies upon the strong intuition that members of the Warsaw Ghetto must not be allowed to perish so that a bullheaded prohibition on private military authorization can be preserved. But, as I have pointed out, a prohibition on private military authorization does not demand that the Warsaw Ghetto residents perish. It may only demand that, when authorizing military force, they do so in an institutionally defensible way. Wars, after all, have the potential to cause widespread death and destruction. To conclude that an individual in the Warsaw Ghetto who chooses to unleash this death and destruction on behalf of residents but without consulting them is acting permissibly—so much so that argumentation is unnecessary—is too quick.

I do not mean to claim yet that such action *would be* impermissible. Much more needs to be said. In particular, a full-fledged theory of public military monopolization is demanded (the ambition of the next two chapters). But, even without such a theory, we may object to the cosmopolitan strategy of defending private military authorization, as the military-claim has not been successfully demonstrated. This claim lacks the intuitive plausibility that is needed for a confident embrace of private military authorization.

2.5 Weaker cosmopolitan critique

Having seen how the stronger cosmopolitan critique falters, we may turn now to the weaker cosmopolitan critique of public military monopolization. The weaker critique implicitly acknowledges that the military-claim alone lacks a degree of intuitive plausibility—that the morally relevant circumstances of individual self-defense are not *the very same* as the morally relevant circumstances of military defense. The extra actors of war, as we just saw, raise difficult questions that tend not to confront cases of individual self-defense, particularly whether individuals must deliberate together about how to proceed. The weaker cosmopolitan critique concedes three constraints to limit military authorization and thereby preserve the plausibility of the military-claim (Fabre 2012: 155): (1) the consent of constituents should be sought; if not obtained, authorizing entities (2) must be convinced that constituents would have consented and (3) must erect institutions to hold authorizing entities accountable after authorization. *Only if* these requirements are met can entities appeal to the individual- and military-claims to justify military authorization.[29]

The three constraints, it should be noted, apply (or must apply) outside of war in order to preserve continuity between the moral grounds of the individual-claim and the moral grounds of the military-claim. A group of individuals, for instance, who are under attack at a supermarket, should, according to Fabre's three constraints, obtain consent from one another about whether and how to proceed; if consent cannot be obtained, those considering defensive action should decide whether others in the supermarket would consent, and these individuals ought to cooperate with legal proceedings after the fact. But none of this is damaging to the weaker cosmopolitan critique. An advocate of this critique is happily willing to apply the three constraints *both* to force short-of-war and to war. Such an embrace lends further plausibility to the military-claim: the circumstances of war *really are* like the circumstances of individual self-defense such that the principles grounding the latter can ground the former.

2.5.1 A consent-based theory of legitimate authority

But, while Fabre presents the three constraints in order to hold strong to a rejection of legitimate authority, a far more plausible interpretation is that these constraints are the *requirements* of legitimate authority. Fulfillment of these constraints *is* the property that entities must possess in order to have legitimate authority.[30] Those who violate the demands of (1), (2), and (3) would seemingly lack justification to authorize war.

The theory may serve to challenge public monopolization if decisions of military authorization governed by requirements of consent are thought to count as private. By contrast, it may support a theory of *public* legitimate authority if the decisions of military authorization governed by the requirements of consent are thought to count as public. Fabre does not elaborate further on the details of her presumed-consent approach—understandably, since it is a caveat aimed to preserve a rejection of legitimate authority rather than a theory of legitimate authority in its own right.

Yet, without further argumentation, it is difficult to see how the decisions that Fabre envisions are not *collective* in the relevant sense. Authorizing entities must be capable of assessing whether consent could be successfully obtained from constituents if attempted and must make some reasonable effort to do so. When consent cannot be obtained, the entity must then seek to promote the interests of *all* constituents by determining whether they would have consented if given the chance. This requires not only a grasp of the threat posed to constituents but also a deep understanding of their risk-tolerance, their just war (or pacifist) leanings, their perceptions of the injustice that provides just cause for war, their attitudes toward the consequences of inaction, and their relations to the threatening party. More importantly, justification for the authorization of war depends upon a *commitment* to honor the interests of constituents. Finally, the authorizing entity must have sufficient confidence in its own institutional capacity to establish a mechanism after (or during) the war for constituents to scrutinize and potentially punish decision makers. Individuals acting alone or groups with *no* institutional capacity at the time of military authorization are unlikely to *justifiably* possess such confidence. None of this, of course, is to suggest that these constraints are enough for proper attribution of legitimate authority. But, as the constraints stand, they alter the character of military authorization so that it is not obviously a private decision (aimed at promoting the individual interests of the authorizing entity). At a minimum, Fabre would have to provide a better account of why these decisions remain private.

Yet, even leaving aside this challenge to the weaker cosmopolitan critique and supposing for the sake of argument that these decisions *do* in fact remain private, there are two deeper worries that confront her account of legitimate authority. First, *why* must consent be sought from constituents? And second, why must consent *only* be sought from constituents (as opposed to others who are also affected by decisions of military authorization)? A theory of legitimate authority ought to be capable of explaining *why* individuals must together possess some set of properties to acquire legitimate authority and *which* individuals count when deciding whether this set of properties is possessed. Lazar (2014: 410–412; forthcoming) resurrects Fabre's position to provide an answer to the first question, but neither Lazar nor Fabre provides an adequate answer to the second.

2.5.2 Why must consent be sought from constituents?

Beginning with the first question, Lazar (2014: 411) argues that, contra Fabre's own professed goals, she provides a strong reason to care about consent, coming "closer than anyone else to explaining precisely why states and cocitizenship matter for the morality of war." To see why, we must understand Fabre's agent-centered theory of self-defense—particularly two features of that theory: the first related to third-party rescue and the second to innocent bystanders. The agent-centered theory of self-defense begins from the proposition that "agents generally have a personal prerogative to confer greater weight on their own projects and goals than on other agents' similar projects" (Fabre 2012: 57). One feature of this theory that supports a role for consent concerns *third-party rescue.* An agent under attack may transfer her power to kill an attacker to a third-party, but, given her personal prerogative, she is also entitled to choose her own death rather than the death of her attacker (Fabre 2012: 62). Transmitting one's power to kill in self-defense to a third-party is dependent, at least to an extent, on one's *consent* to that transmission.

A second feature of the theory that supports a role for consent concerns *innocent bystanders*. An agent may kill an innocent bystander in self-defense when doing so is necessary to preserve her own life (Fabre 2012: 57). *But,* the agent may not always transfer her power to kill innocent bystanders to (disinterested) third-parties, because the third-party may lack the personal prerogative that justifies killing innocent bystanders in the first place; such transfer will be permissible (in some instances) only when the third-party is defending jointly held rights with the agent and only when both *consent* to defend these rights (Fabre 2012: 68). Consent will be required whenever the agent-neutral reasons that might justify killing innocent bystanders (such as "not living in a world in which wrongful attackers are allowed a free rein") are insufficiently robust (Fabre 2012: 62–63, 68).

As should be clear, both features of Fabre's agent-centered theory of self-defense provide reasons for why consent matters in the authorization of war. The first feature would insist that authorizing entities obtain consent from their constituents, because constituents have the right to forego war.[31] The second feature would insist upon consent, because as Lazar (2014: 410) reminds us and Fabre (2012: 89) herself acknowledges, war *always* involves the collateral killing of innocent bystanders, and the jointly held rights that permit such collateral killing may demand that rights-possessors *choose* to defend one another. Just as an individual has the prerogative to forego exercising her individual right to defense, so too does a group of individuals have the right to forego exercising their jointly held rights to defense. If such jointly held rights—shared by the authorizing entity and its constituents—are to permit the authorizing entity to undertake killing on behalf of constituents that will result in the collateral death of innocent bystanders, then the entity ought to obtain the consent of constituents.

While an agent-centered theory of self-defense provides one set of reasons to favor constituent-consent in military authorization, Lazar (forthcoming: 10–15) points to associative duties as another. He argues that associative duties (to fellow constituents, loved ones, friends, etc.) may function just like agent-centered prerogatives to help justify killing innocent bystanders. In the same way that consent may be required by an individual to transmit an agent-centered prerogative to defend herself, consent may be required by an individual to transmit an associative duty to defend her loved ones. Both agent-centered prerogatives and associative duties yield agent-relative reasons for action, and "the reason would not be agent-relative" if an entity could "act on another's agent-relative reasons without being authorised to do so" (Lazar forthcoming: 11).[32]

2.5.3   Why must consent be sought only from constituents?

Without delving into the merits of agent-centered self-defense or theories of associative duties—both controversial in their own rights—neither Fabre nor

Lazar provide an answer to the second question posed above: why must consent *only* be sought from constituents? Lazar is well aware of this limitation. He expressly disavows an interest in providing a full-fledged theory of legitimate authority and instead aims to explain why consent (broadly understood) matters in a certain subset of wars: namely those *authorized on behalf of a community* (i.e., those with constituents). But, even in such wars, this approach, and indeed Fabre's weaker cosmopolitan challenge, are incomplete, as they do not defensibly account for *whose* consent matters.

Let me explain by way of two examples. Both underscore why a focus on constituents alone is incapable of delivering a complete theory of legitimate authority. First, consider a state with three ethnic groups—A, B, and C—each occupying a parcel of land within the state. Group A has long oppressed Groups B and C, and the military of Group A occupies the lands of Groups B and C. The leaders of Group B, fed up, seek and obtain the direct consent of every single constituent in Group B to authorize military force against Group A. The constituents are aware that such authorization carries a significant risk of reprisal by Group A against both Group B and Group C (since Group A will not be able to discern who staged the attack). Given Group A's war-fighting in the past, Group B is convinced that these reprisals will not be catastrophic (though many civilians will be killed). Group B is thus confident, particularly given the severity of oppression by Group A, that the authorization of war against Group A will be proportionate. Aware of the risks, Group B decides to authorize war against Group A. They do so on their own behalf, not on behalf of Group C. Indeed, they have strained relations with Group C, and, in any case, a foreign entity, Group D, will immediately occupy the territory of Group C if Group A is toppled. Luckily for Group B, Group D has no interest in occupying their territory.

Is it correct to say that the leaders of Group B have *no* responsibility to *attempt* consultation with the leaders or constituents of Group C before authorizing military force? Fabre and Lazar would insist that the leaders of Group B must attempt to consult *those on behalf of whom* they authorize war, namely the constituents of Group B. But Group B is *not* authorizing war on behalf of Group C; indeed, Group C stands to enjoy no benefit if Group A is defeated. Fabre and Lazar thus have no grounds to insist upon consultation with Group C according to the constraints that they place on military authorization. Is it justifiable for the leaders of Group B to authorize war having made *no* effort to consult with the members of Group C? The thought that no consultation is required should be particularly worrying for a cosmopolitan like Fabre who maintains that borders are (mostly) irrelevant for the assignment of rights and responsibilities.

Now consider a second (similar) example. A private protective agency (PPA_B) in Group B, who represents 1,000 individuals from the group's 100,000 total members, is fed up with the occupation of Group B's territory by Group A. The board of PPA_B seeks and obtains the direct consent of all 1,000 of its clients to authorize military force against Group A. These clients are aware that such authorization carries a significant risk of reprisal by Group A against both

Groups B and C, though they are confident that their authorization of war will remain proportionate since the reprisals are unlikely to be catastrophic (though many civilians will be killed). Despite the risks, the board of PPA_B decides to authorize war against Group A. They do so on their own behalf, not on behalf of other members of Groups B or C. Indeed, the occupation has had far more devastating effects on the clients of PPA_B than anyone else in these groups, leading to disgruntled relations between the clients of PPA_B and the members of Groups B and C. In any case, Group D will immediately occupy the territory of Groups B and C if Group A is toppled. Luckily for the clients of PPA_B, Group D will treat them far more favorably than Group A did, though Group D will treat Group B and Group C just the same as Group A treated them.

Again, Fabre and Lazar have no way of explaining why perhaps PPA_B ought not to authorize military force without consulting the members of Groups B and C. PPA_B has the direct consent of all individuals on behalf of whom it authorizes force. The war will be proportionate. Yet, PPA_B has imposed substantial risks upon members of Group B and Group C without their consultation (and with no hope that the risks will yield some benefit to them). Is this risk-imposition justifiable? The next two chapters will argue that it is not. The important point for now is simply that a theory of legitimate authority must account for more than simply the duties that an authorizing entity owes to its constituents.[33]

2.6 Conclusion

To summarize then, we must conclude that the stronger and weaker cosmopolitan defenses of private military authorization have faltered and that the public monopolization of military authorization remains unchallenged. The stronger case relied upon questionable intuitions, and the weaker case, which aimed to shore up these intuitions, collapsed into a theory of legitimate authority. The resulting theory could not establish a private right of military authorization since the stringent constraints placed upon decision-making left doubt about whether decision-making remained private in the relevant sense. More importantly, the theory itself could not explain why authorizing entities ought to be concerned only with the consent of their constituents.

Notes

1 See Langan (1984) for a helpful review of Augustine's just war theory, which includes the requirement of "supreme authority" to initiate war. On Aquinas' view of authority to wage war, see Johnson (1999: 44–51; 2003: 7–13).
2 The other *jus ad bellum* requirements (likelihood of success, proportionality, and last resort), by contrast, were initially "prudential tests" for political decision-making borrowed from Roman practice by Augustine (Johnson 1999: 42).
3 For this point about the relative neglect of legitimate authority, see also Orend (2000: 192), Held (2005: 175), Finlay (2010: 287), Fabre (2012: 142), Schwenkenbecher (2013: 161), Reitberger (2013: 65), Parry (2015: 176). Note that while contemporary theorists have tended to emphasize just cause above other *jus ad bellum conditions*, just war theorists in the past often emphasized legitimate authority (Johnson 2003: 7).

Aquinas (2003 [1266–73]: 165) insisted that *the first requirement* of just war was "that the ruler at whose command the war is to be waged have the lawful authority to do so."

4 See Parry (2015: 177, 185) for the language of legitimate authority as the "possession of a certain set of properties."

5 The older terms of "sovereign authority," "proper authority," and "competent authority" will not be used.

6 The terms "monopolization over military authorization" and "monopolization" will be used interchangeably.

7 Fabre (2012: 142–156) is a revised version of Fabre (2008). I will cite the former version except in those instances where the two diverge, and relevant material in the latter is absent from the former.

8 Fabre (2012: 142) reminds us that judgment about the war and the subsequent action attending to this judgment are not necessarily conjoined: "very few philosophers take the view that no one except legitimate war-wagers can judge whether the war is just." But the point is that, the two tasks together—judgment and action—are ordinarily limited to a certain kind of actor. And the question with which we are concerned is: what kind of actor should that be?

9 In referring to authority as a claim-right and a liberty-right, I follow Buchanan (2002: 691–692; 2004: 237) and Raz (1985: 5–6). But keep in mind that Buchanan and Raz defend conceptions of *political authority* and are interested in claim-rights that entail obedience to the dictates of a state. My conception of *legitimate authority* in war entails no claim-rights of obedience but rather (as will become clear in a moment) claim-rights of forbearance. See Lazar (forthcoming: 2) for the point that the two literatures—on political authority and legitimate authority in war—have largely remained separate from one another. My attempt here is not to unite these literatures, only to borrow a definitional point from the former to facilitate discussion of the latter.

10 See Christiano (2013: 7) for a similar point in his discussion of authority.

11 See also Finlay (2013: 152).

12 See Johnson (1999: 42–43) and Reitberger (2013: 67) for reference to the "deontological" requirement of legitimate authority—which they understand to mean that legitimate authority is a *necessary condition* for the justified authorization of war. Note how my account of legitimate authority is deontological in a different sense. It is deontological in the sense of being rights-based. A failure to possess legitimate authority when authorizing war is wrong because it violates the rights of others.

13 These sorts of emergency circumstances will be considered in Section 2.4.2 and Section 4.6.

14 An interest-based theory of rights is elaborated in Section 3.4.1, and its relationship to will-theories is elucidated.

15 It is true, of course, that actors who possess legitimate authority may be justified in authorizing war against those who lack legitimate authority. But most preventive measures aimed at keeping individuals and groups who lack legitimate authority from authorizing war entail intelligence-gathering, policing, imprisonment, and so on, not war itself.

16 Notice that, because the justified monopolization of military authorization includes a liberty-right and claim-right, it is also a type of authority (on my account). I will refer to the "authority of entities to monopolize military force" and "the right of entities to monopolize military force" synonymously.

17 This point must be made since, as the global spate of civil wars reminds us, military force is not always authorized against foreigners, and thus the right of military monopolization may contain non-federative as well as federative elements.

18 By non-military force, I mean to include all forms of force short-of-war. Though the military may be involved in the use of force short-of-war, military force, for the

purpose of this book, will refer only to war, while non-military force will refer to all forms of force besides war. See Section 1.4.

19 As we will see, other prominent cosmopolitan thinkers, including Caney (2005) and Rodin (2002), do not reject the requirement of legitimate authority but instead embrace a more global monopolization of military force.

20 Fabre cites Vitoria and Grotius as early advocates of her position. Vitoria, for instance, writes that "any person, even a private citizen, may declare war and wage a defensive war" (in Fabre 2012: 146–147).

21 For a similar argument, see Reitberger (2013: 72):

> if it can be justifiable to use lethal force to defend yourself and others against unjustified armed attack, it should be possible to justify large-scale collective self-defense up to and including the scale of war as well, regardless of authorization or legal status.

22 Fabre is largely silent on the role of consent in her stronger cosmopolitan challenge of legitimate authority. If anything though, she is skeptical of its value, claiming: (1) that governments often lack consent from constituents to authorize war but should not be prohibited from war making (Fabre 2008: 973); (2) that, if representative decision-making stands in for consent, authorization by representatives "is no reliable indication that their constituents ... actually approve that decision" (Fabre 2008: 973); and (3) that the requirement of consent would be "impossibly demanding," as it would render certain just wars unjust contrary to our intuitions (e.g., had Roosevelt bypassed Congress to authorize war after Pearl Harbor). All of these claims are consistent with the proposition that consent should at least be sought prior to authorization (as her weaker claim demands), but Fabre does not limit her stronger cosmopolitan critique in this way.

23 I do not consider an objection here raised by Anna Stilz (2014: 324–325), who argues that just because some individual right is valuable does not entail that the rights-possessor be the one who does the enforcing: only if "an institution is unavailable or unreliable" does the right permit an individual "exceptional permission to defend his own rights." Fabre's weaker cosmopolitan critique, as she herself (2014: 408) points out in response to Stilz, endorses this position, and so I will defer assessment to Section 2.5.

24 These weapons, of course, might also be off limits for *public* possession on the same grounds. However I leave aside this question here.

25 The term "dual-use" is typically used to describe facilities that are targeted in war with both civilian and military functions. See, for example, Shue and Wippman (2002). However the term should also be suitable here.

26 This analysis of the first retort is even stronger when we remember that the agent who authorizes military force need not be the agent who supplies military force. A private authorizing entity does not have to *itself* possess the weapons and training that are needed to prosecute war. So long as *someone* possesses these weapons and training and is willing to use them on behalf of the private authorizing entity, then the private authorizing entity may be capable of authorizing war. Forces may even be hired from abroad, and thus the existence of a single territory with an ineffectual ban over single-use weapons and training or a single territory where dual-use weapons and training are available is enough to demand further analysis of private military authorization.

27 Rodin proposes this example not to endorse the view that private military authorization is sometimes permissible but to underscore the point that *group-defense*, in contrast to national-defense, is sometimes permissible on the grounds of self-defense.

28 One final point is worth noting. Even if the institutional worry just raised were set aside and all Warsaw Ghetto residents were treated as private actors, at least one further institutional question remains. How should governments around the world

respond to an uprising by Warsaw Ghetto residents? Should these governments, in the name of public military monopolization, seek to thwart the private authorization of military force? Or should they instead remain neutral or offer assistance?

29 An internal critique of this move by Fabre, which will not be pursued further (though see Section 4.6 for more on emergency authorization), is that it renders her central cosmopolitan position vulnerable to arguments from emergency authorization—to precisely the sort of arguments that are meant to lend her central cosmopolitan position intuitive plausibility. Suppose that an authorizing entity is not convinced that constituents would consent to military authorization (they really do not know how constituents would decide) or cannot erect institutions to ensure accountability after military authorization (and are aware, prior to the war, of their likely inability to do so). Are there *no* circumstances in which individuals may still be justified in waging war without meeting these criteria? If there are no such circumstances, then Fabre must explain why (as we will see below). If there *are* such circumstances, then Fabre's weaker cosmopolitan critique is open to the criticism that she employs against proponents of legitimate authority: individuals have a right to defend themselves, and failure to belong to a group that meets the three criteria of the weaker cosmopolitan critique cannot be grounds for denying individuals this right (Fabre 2012: 145).

30 See Lazar (2014: 411) who rightly notes that while Fabre "presents her presumed-consent account of legitimate authority as an alternative to a theory of legitimate authority," it is really just a "version of such a theory." See also Schwenkenbecher (2013: 167).

31 See also McMahan (2010: 48), McPherson (2007: 544), and Finlay (2010: 290) on this point. McPherson (2007: 544) provides the helpful example of India's anti-colonial struggle against the British, where it would have been wrong "to disregard the Indian people's ethical and strategic commitment to pursuing independence through nonviolent resistance under Gandhi" by authorizing force on their behalf. Likewise, Finlay (2010: 294) points to Buddhists and Quakers who value "their being and their capacity for moral choice based on the ability to realize pacifist commitments these entail" such that saving "them from unjust aggression using lethal violence would be self-contradictory."

32 Lazar is clear that, according to his extension of Fabre's theory, consent—which he understands broadly to include direct consent, presumptive consent, and democratic choice—is not a *requirement* for justified military authorization. Rather, it is one consideration that affects a war's overall proportionality. Indeed, Lazar (forthcoming: 4–10) identifies several further reasons (beyond those already mentioned) why consent may alter this calculus of proportionality: (1) initiating war without the endorsement of constituents (while using their resources) disrespects them by treating their judgments as inferior to those of the authorizing entity; (2) war is costly for constituents, and such costs should not be imposed without seeking their endorsement; (3) war entails enormous moral risk for constituents—namely that the war will turn out to be unjustified and that individuals will be wrongfully killed—and constituents have a right (and a duty) to ensure that moral risks undertaken in their name and with their resources are defensible; (4) the duties that individuals defend in war (associative and otherwise) may cease to apply if beneficiaries release these individuals from their duties; and (5) in wars of political independence, endorsement by constituents preserves respect for the cause that motivates war (namely independence). Each provides a further answer to the first question identified above confronting Fabre's theory of legitimate authority: *why* must consent be sought from constituents?

33 The same objection may be leveled against other prominent approaches to legitimate authority: Finlay (2010: 288), for example, is interested in the legitimate authority of "non-state entities when they claim to *act on behalf* of the victims of rights violations and political injustice." McPherson (2007: 541) sees the distinctive wrong of terrorism

as being "the resort to political violence without adequate license from a people *on whose behalf* the violence is purportedly undertaken." Schwenkenbecher (2013: 164) insists that "an agent fighting a political injustice with violent means usually has moral authority to employ violence for this end if she has explicit approval of the people *on whose behalf* she acts." (My emphasis is added.)

References

Applbaum, Arthur Isak (2007). "Forcing a People to Be Free," *Philosophy and Public Affairs*, Vol. 35, No. 4, pp. 359–400.

Aquinas, Thomas (2003 [1266–73]). "Excerpts from *Summa Theologiae*," in William P. Baumgarth and Richard J. Regan (eds), *On Law, Morality, and Politics*, second edition (Indianapolis, IN: Hackett Publishing Company).

Buchanan, Allen (2002). "Political Legitimacy and Democracy," *Ethics*, Vol. 112, No. 4, pp. 689–719.

Buchanan, Allen (2004). *Justice, Legitimacy and Self-Determination: Moral Foundations for International Law* (Oxford: Oxford University Press).

Caney, Simon (2005). *Justice Beyond Borders: A Global Political Theory* (Oxford: Oxford University Press).

Christiano, Tom (2013). "Authority," in Edward N. Zalta (ed), *The Stanford Encyclopedia of Philosophy* (Spring 2013 Edition). Available at http://plato.stanford.edu/entries/authority/ (accessed December 28, 2015).

Dworkin, Ronald (1984). "Rights as Trumps," in Jeremy Waldron, *Theories of Rights* (Oxford: Oxford University Press).

Fabre, Cécile (2008). "Cosmopolitanism, Just War Theory, and Legitimate Authority," *International Affairs*, Vol. 84, No. 5, pp. 963–976.

Fabre, Cécile (2012). *Cosmopolitan War* (Oxford: Oxford University Press).

Fabre, Cécile (2014). "Rights, Justice, and War: A Reply," *Legal Theory*, Vol. 33, No. 3, pp. 391–425.

Finlay, Christopher J. (2010). "Legitimacy and Non-State Political Violence," *The Journal of Political Philosophy*, Vol. 18, No. 3, pp. 287–312.

Finlay, Christopher J. (2013). "Fairness and Liability in the Just War: Combatants, Non-Combatants, and Lawful Irregulars," *Political Studies*, Vol. 61, No. 1, pp. 142–160.

Gross, Michael L. (2013). "Just War and Guerilla War," in Anthony Lang, Cian O'Driscoll, and John Williams (eds), *Just War: Authority, Tradition, and Practice* (Washington, DC: Georgetown University Press), pp. 213–229.

Held, Virginia (2005). "Legitimate Authority in Non-State Groups Using Violence," *Journal of Social Philosophy*, Vol. 36, No. 2, pp. 175–193.

Hohfeld, Wesley (1919). *Fundamental Legal Conceptions as Applied in Judicial Reasoning* (New Haven: Yale University Press).

Johnson, James Turner (1999). *Morality and Contemporary Warfare* (New Haven: Yale University Press).

Johnson, James Turner (2003). "Aquinas and Luther on War and Peace: Sovereign Authority and the Use of Armed Force," *The Journal of Religious Ethics*, Vol. 31, No. 1, pp. 3–20.

LaFollette, Hugh (2000). "Gun Control," *Ethics*, Vol. 110, No. 2, pp. 263–281.

Langan, John (1984). "The Elements of St. Augustine's Just War Theory," *The Journal of Religious Ethics*, Vol. 12, No. 1, pp. 19–38.

Lazar, Seth (2014). "Review of *Cosmopolitan War*," *Ethics*, Vol. 124, No. 2, pp. 406–412.

Lazar, Seth (forthcoming). "Authorisation and the Morality of War," *Australasian Journal of Philosophy*, pp. 1–19.

Locke, John (2002 [1690]). *The Second Treatise of Civil Government* (Austin: Liberty Library). Available at www.constitution.org/jl/2ndtreat.htm (accessed December 28, 2015).

McMahan, Jeff (2010). "Humanitarian Intervention, Consent, and Proportionality," in N. Ann Davis, Richard Keshen, and Jeff McMahan (eds), *Ethics and Humanity: Themes from the Philosophy of Jonathan Glover* (Oxford: Oxford University Press), pp. 44–72.

McPherson, Lionel (2007). "Is Terrorism Distinctively Wrong?," *Ethics*, Vol. 117, No. 3, pp. 524–546.

Orend, Brian (2000). *War and International Justice: A Kantian Perspective* (Waterloo: Wilfrid Laurier University Press).

Parry, Jonathan (2015). "Just War Theory, Legitimate Authority, and Irregular Belligerency," *Philosophia*, Vol. 43, No. 1, pp. 175–196.

Pattison, James (2014). *The Morality of Private War* (Oxford: Oxford University Press).

Raz, Joseph (1985). "Authority and Justification," *Philosophy and Public Affairs*, Vol. 14, No. 1, pp. 3–29.

Reitberger, Magnus (2013). "License to Kill: Is Legitimate Authority a Requirement for Just War?," *International Theory*, Vol. 5, No. 1, pp. 64–93.

Rodin, David (2002). *War and Self-Defense* (Oxford: Oxford University Press).

Schwenkenbecher, Anne (2013). "Rethinking Legitimate Authority," in Fritz Allhoff, Nicholas G. Evans, and Adam Henschke (eds), *Routledge Handbook of Ethics and War: Just War Theory in the Twenty-First Century* (Oxford: Routledge), pp. 161–170.

Shue, Henry and David Wippman (2002). "Limiting Attacks on Dual-Use Facilities Performing Indispensable Civilian Functions," *Cornell International Law Journal*, Vol. 35, No. 3, pp. 559–579.

Stilz, Anna (2014). "Authority, Self-Determination, and Community in *Cosmopolitan War*," *Legal Theory*, Vol. 33, No. 3, pp. 309–335.

Thompson, Janna (2005). "Terrorism, Morality, and Right Authority," in Georg Meggle (ed), *Ethics of Terrorism & Counter-Terrorism* (Frankfurt: Ontos Verlag), pp. 151–160.

Weber, Max (2004 [1919]). "Politics as a Vocation," in Rodney Livingstone (trns), *The Vocation Lectures* (Indianapolis: Hacket Publishing).

Wenar, Leif (2015). "Rights," in Edward N. Zalta (ed), *The Stanford Encyclopedia of Philosophy* (Fall 2015 Edition). Available at http://plato.stanford.edu/entries/rights/ (accessed December 28, 2015).

3 All affected fundamental interests

Part I has thus far challenged one important defense of private military authorization: that of Cécile Fabre. The next two chapters offer positive cases for public monopolization and public military authorization. I aim to show first that public entities must prevent private entities from *authorizing* military force. In light of this argument, A_1 and A_2 must be rejected. I then seek to show that public monopolizing entities have a responsibility to authorize military force when the security of those who have been disarmed is threatened. The upshot of the second argument is that A_0, the non-provision of military force, must be rejected. If my analysis in these two chapters is sound and public institutions must monopolize the authorization of military force when needed, then only three arrangements will remain as candidates for our potential endorsement: A_3, A_4, and A_5.

This chapter presents the first of a two-part case for public military monopolization. The position that I defend for public monopolization is referred to as the *risk-prevention argument*. According to the risk-prevention argument, private military authorization is unjustifiable and public military monopolization justifiable (and indeed obligatory), because private military authorization would impose considerable risks on individuals who lack a say in authorization (and we ought not to impose considerable risks on individuals who lack such a say). Individuals who are affected by decisions of military authorization must have an opportunity to voice their input.

Which individuals, we may wonder, ought to have such an opportunity? One set of individuals who should *not* have this opportunity, I argue, are the enemy combatants under attack by private actors who have forfeited their freedom from substantial risk-imposition. But two other sets of individuals, I seek to show, both of whom are subjected to risk by private decisions of military authorization, *do* deserve a say. One set (perhaps counter-intuitively) are members of the community against whom private force is authorized who have *not* forfeited their freedom from substantial risk-imposition and who lack a meaningful voice in their own government. The second set are members of the international community—especially those who reside in geographic locations that are close in proximity to or may *become* close in proximity to the theater of hostilities—who have not forfeited their freedom from substantial risk-imposition. According to my analysis, private actors who authorize military force without the input of these groups subject them to unjustifiable risks. Consequently, both groups may be entitled to participate in decisions of military authorization.

One important class of individuals who are included among the second group just identified are the *affiliates* of private authorizing entities: the co-nationals, coreligionists, and others who may become targets of retaliation on account of their shared group membership with, or proximity to, private authorizing entities. The unjustifiable risks imposed by private actors authorizing military force may be imposed *directly* (when *their* military harms those with immunity) or *indirectly* (for instance, when they cause the enemy to harm those with immunity). It is the latter variety of risk—indirect risk—that typically plagues affiliates. A paradigm example was the Al-Qaeda authorization of military force against the US, which caused the US to retaliate against Afghanistan. Al-Qaeda, on my view, subjected its affiliates (the Afghani people) to unjustifiable risks by authorizing force against the US.[1] In light of these indirect risks, the affiliates of private authorizing entities must be included in private decisions of military authorization. And because, according to the risk-prevention argument, such inclusion must be collectively *ensured*, the ultimate decision of military authorization ought to be withdrawn from the private sector and reserved for public discharge (in other words, it should be publicly monopolized).

Versions of the risk-prevention argument have been sketched by several political theorists, some registering their support for its conclusion (McMahan 2010a: 47–54; Pattison 2010: 142–144; Lazar forthcoming: 4–5) and others expressing opposition (Fabre 2012: 150–156). Though the versions that other theorists have sketched provide a helpful starting point for analysis, they are either too narrow for our examination—speaking, for instance, only of the need for public authorization in the context of humanitarian intervention—or are ineffectively framed. I will construct a version of the risk-prevention argument that includes the full range of wars under the reach of its conclusion and attempts to shore up the weaknesses that challenge alternative versions. In developing my argument, I borrow insights from a variety of literatures, including work on the boundary problem in democratic theory, risk-assessment in decision theory, and the precautionary principle in environmental theory. The argument, I think, represents a promising but insufficiently charted approach to the monopolization of military authorization.

The risk-prevention argument for public military monopolization may be captured in two broad premises: the *all affected fundamental interests premise* and the *risk-imposition of war premise*.

P$_1$ *All Affected Fundamental Interests*: A decision ought to be withdrawn from the private sector and reserved for public discharge when one course of action under consideration imposes risk, above some threshold level, to the fundamental interests (e.g., physical security) of enough individuals. Fundamental interests are interests that are sufficiently weighty so as to be protected by rights.

P₂ *Risk-Imposition of War*: The private decision to authorize military force imposes risk (directly and indirectly), above an acceptable threshold level, to the fundamental interests of enough individuals. Specifically, the private decision to authorize military force imposes such risks to our fundamental interests in physical security.

This chapter defends P_1, and Chapter 4 defends P_2. If these two premises are correct, then the decision to authorize military force ought to be withdrawn from the private sector and reserved for public discharge.[2]

The current chapter proceeds as follows. In the first section, I briefly consider and dismiss five prominent arguments for the public monopolization of military authorization.[3] In the second section, I introduce the risk-prevention argument by presenting versions that have been put forward by Jeff McMahan, James Pattison, Cécile Fabre, and Seth Lazar. I consider these versions to extract several compelling claims that I later endorse. With these versions of the risk-prevention argument and their limitations in mind, I take up P_1. In the third section, I elucidate the principle from which P_1 is derived—the all affected interests principle. The fourth section specifies and defends four sub-premises that together constitute P_1. Once my defense of these sub-premises and hence P_1 is finalized, I move on to consider P_2 in Chapter 4.

3.1 Five just war arguments for public monopolization

The first prominent defense of public monopolization that we should consider is offered by Simon Caney (2005: 206), who follows Aquinas in arguing that private entities, at least in the current international system, must be prohibited from authorizing military force, because they have an available alternative to rectify their grievances: the state. As Aquinas (1988 [1266–73]: 221) writes, "it is not the business of a private individual to declare war, because he can seek redress of his rights from the tribunal of his superior."[4] The problem with this position is that it will not suffice when individuals lack access to legal recourse, for example when their state violates or fails to prevent the violation of their rights (Fabre 2012: 150; Reitberger 2013: 80, 82). Moreover, the position fails to identify *why* deference must be paid to public bodies in those cases where individuals do possess legal recourse for their grievances. Why must private individuals avail themselves of legal channels and appeal to the tribunal of their superiors rather than simply authorize military force as they see fit?

A second argument is that private entities must not be permitted to authorize military force, because individuals reasonably disagree about when and how military force should be authorized—and private military authorization would subvert the fair resolution of such disagreement (Caney 2005: 207; Stilz 2014: 329–335). As Stilz (2014: 332) argues following Kant: "When one private individual forces another to submit to his judgments about justice, he undermines the relation of reciprocal equality in which individuals ought to stand, by subjecting this other person to his private power." The only fair solution is to establish just institutions that enable resolution of reasonable disagreement. While I believe (and will argue) that reasonable disagreement over the authorization of war *should* factor into our rejection of private military authorization, I do not believe that it can alone constitute a rejection of private military authorization. Countless decisions in a society are characterized by reasonable disagreement, and yet we do not prevent private actors from making all of these decisions. For instance, people disagree about which religion is most credible, but we do not insist that decisions about religion be withdrawn from the private sector and made by public institutions. If reasonable disagreement is to motivate such a withdrawal in the context of military authorization but not in the context of religious belief, a further argument is needed to show why reasonable disagreement in the context of military authorization is the kind of reasonable disagreement that demands public monopolization. I have no doubt that it does; indeed, the risk-prevention argument provides the further argumentation needed to demonstrate as much. But reasonable disagreement is not alone sufficient to reject private military authorization.

A third argument against the private authorization of military force is offered by James Pattison (2008: 150).[5] Pattison presents two formulations of the argument, a stronger formulation and a weaker formulation. According to the stronger formulation, public monopolization over military authorization is *necessary* for the regulation of war. He writes: in "restricting which particular agents can use force," monopolization "makes it possible to establish legal and political instruments that govern and regulate warfare," which in turn "help to provide a common framework to reduce the horrors, and frequency, of military force" (Pattison 2008: 150). To the credit of this position, it is certainly true that contemporary international humanitarian law depends upon the public monopolization of military authorization. But the existence of legal and political instruments that govern warfare is not *necessarily dependent* upon the public monopolization of military authorization. Under international commercial law, another realm of regulation, private entities engage in business exchanges with one another and with states. But these business exchanges are nevertheless effectively regulated by mechanisms like the World Trade Organization (WTO). The WTO does not require the public monopolization of trade in order to function. It would seem that similar regulatory schemes could apply to war if private entities were permitted to authorize military force against one another and against states. Laws governing private transactions are not, in any obvious way, necessarily inapplicable to war. Seemingly then, the public monopolization of military authorization is not defensible on the grounds that it is necessary for the regulation of war.

A weaker formulation of this argument is that the public monopolization of military authorization is *advantageous*, not necessary, for the regulation of war (Pattison 2008: 150; Coates 1997: 125).[6] As Pattison (2014: 169) acknowledges, however, it is not clear that a world of public actors who alone possess legitimate authority *is* more peaceful than alternatives. We need only consider "the brutal wars, the mass devastation, the A-bombing that our State-ridden world has suffered for centuries"—including the Dresdens and Hiroshimas (Rothbard 1973: 226).[7] Second, and the first point notwithstanding, even if the public monopolization of military authorization would make the regulation of war easier, the enhanced ease could not sustain an argument for public monopolization. Imagine an international system where only states whose capital city begins with the letter A were permitted to authorize military force. This norm, which limits the number of actors who may justifiably authorize military force, could make the regulation of war far easier than alternatives. Yet, if a state whose capital city started with the letter B were attacked, the norm would not provide a good reason for this state to forego authorizing military force. The simple point is that a norm may be questionable even if it facilitates enhanced regulatory efficacy. In the example just given, the norm may violate the rights of states whose capital cities start with B when these states are forced to suffer their demise. To all but the committed consequentialist, a defense of public monopolization that relies upon the promotion of enhanced regulatory efficacy will be unsatisfying.

A fourth argument for public monopolization is that it is needed to preserve democratic control over the authorization of military force (Pattison 2008: 153; Schwenkenbecher 2013: 161; Valls 2000: 71; Held 2005: 187; Thompson 2005: 93; McPherson 2007: 544–545). The argument for democratic control can be made on the *instrumental* grounds of promoting peace and the *intrinsic* grounds of self-governance (McPherson 2007: 544–545; Pattison 2008: 153). I will leave aside the former justification, as I believe it is vulnerable to the same deontological challenge that was just raised against the argument from enhanced regulatory efficacy. The difficulty with the latter justification is, first of all, that the private authorization of military force is not obviously at odds with the intrinsic values of self-government and individual autonomy. To see this, consider a contrasting example in which a private decision *does* seem to be at odds with these values. When a state is set to make a decision as a collective on whether to build highways, and a private actor supplants public judgment with her own judgment, then the values of self-government and individual autonomy are undoubtedly undermined. Citizens are excluded from deciding whether or not *their* state will build more highways. But when a private individual authorizes military force on her own accord (and not on behalf of her state), she is not preventing citizens from deciding whether or not *their* state is going to authorize war.[8] Citizens may still decide for themselves whether or not to authorize war.[9]

Moreover, even supposing *arguendo* that the values of self-government and individual autonomy *are* undermined by the private authorization of military force, still it is not clear that these values are sufficient to ground public military monopolization.[10] Consider an example proposed by Pattison in which a government makes undemocratic decisions but always promotes its citizens' interests. Pattison (2010: 138) writes correctly, I think, that while "such a government would not be *that* morally objectionable because it would be promoting its citizens', something morally important is still missing."[11] This morally important, missing piece is not one that Pattison believes can alone ground public military monopolization. Rather, it is pitched as one consideration among several that weigh against the justifiability of private military authorization. But the argument from democratic control may nevertheless be assessed for its stand-alone plausibility (particularly since all other arguments have so far been challenged). When considered in this way, it seems inadequate to ground public military monopolization.

The fifth and final argument to take up against private military authorization concerns distributive justice. According to this argument, only through the public monopolization of military authorization can a society ensure that military protection will be provided equally, or even sufficiently, to all members.[12] In many societies today, like the US, where economic and social inequalities are colossal, one good that is truly enjoyed more or less equally by all citizens is military protection. The wealthiest and poorest citizens are protected by the US military roughly to the same extent. When military authorization is required against some threat, the wealthy and poor will either be protected together or not at all.[13]

While this argument is compelling on its face, it suffers from a simple problem: societies need not monopolize military force to ensure the proper distribution of military protection. They may ensure such a distribution by issuing vouchers to be redeemed with private protective agencies, which are permitted to authorize military force (and which, we may suppose, are not allowed to turn clients away). These vouchers may serve sufficientarian conceptions of distributive justice: just as food stamps are issued to feed the poor, protection stamps may be issued to safeguard the vulnerable. Vouchers may also be distributed to neutralize the effects of bad brute luck—to those who live in dangerous areas and those who are naturally fearful of foreign attack (if "expensive tastes" are accommodated). More generally, vouchers may be conjoined with the redistribution of other goods (including health care, education, and so on) to ensure that one's total bundle of goods will correspond to the demands of *any* favored distributive principle, whether sufficientarianism, luck egalitarianism, or something else. The government, to provide extra assurance, may prohibit citizens from receiving top-up protection beyond the distributive principals that are enshrined in vouchers; it may punish private protective agencies that fail to promote these distributive principles; and, when the principles are violated, it may serve as a safety-net authorizing entity for individuals who receive inadequate protection. The government may do all of this without *monopolizing* the authorization of war.

3.2 The risk-prevention argument

The risk-prevention argument offers an alternative to these five rationales for the monopolization of military authorization. McMahan (2010a: 47–54) and Pattison (2010: 142–144) each present versions of the risk-prevention argument in the context of humanitarian intervention. In particular, both claim that, because humanitarian interventions impose risks upon the intended beneficiaries,[14] these intended beneficiaries must have a say in military authorization.[15] McMahan (2010a: 50) writes: humanitarian intervention "can seldom promise rescue without also endangering its beneficiaries," and we must not "expose people to the risk of such harm in the absence of compelling evidence that they are willing to accept that risk for the sake of the promised benefits." If the individuals on behalf of whom an intervention is authorized genuinely do not *want* the intervention to proceed, then, on the view of McMahan and Pattison, it ought not to proceed.[16]

We should note that whereas McMahan requires interveners to seek the *consent* of intended beneficiaries, Pattison demands that interveners attempt to *represent* them. Yet, while this distinction may seem consequential, it belies an essential congruity between the two views. McMahan acknowledges that the acquisition of consent from all who are subjected to risk, which would be ideal, is nearly impossible given the time pressures of humanitarian intervention. Moreover, even if it were possible, those subjected to risk are almost never in unanimous agreement (or unanimous opposition) to the proposed intervention. For these reasons, McMahan (2010a: 47) takes consent to be "something more like a widespread or general desire for intervention." This acknowledgment thus renders the consent that McMahan advocates almost identical to the representation that Pattison advocates. As Pattison (2010: 140) insists, representation requires that interveners "establish the opinions" of intended beneficiaries by first attempting to "obtain direct access" to them and, if that is not possible, using "secondary sources or indicators of these citizens' opinions, provided, for instance, by intermediaries." For both McMahan and Pattison, would-be interveners must seek to ascertain the preferences of intended beneficiaries, whether through consultation or some other means, and act upon these preferences.

In at least one crucial respect, however, Pattison's version of the risk-prevention argument is superior to McMahan's version. Pattison maintains that bystanders who are subjected to risk in an intervention, and not just intended beneficiaries, must be included in the decision of authorization. To expand upon the relevant group of individuals who are granted input, Pattison borrows a distinction from Fernando Tesón (2003: 106) between the accomplices (and collaborators), bystanders, and victims of a humanitarian crisis.[17] Whereas McMahan restricts his analysis to victims (who I have been calling intended beneficiaries[18]), Pattison endorses consultation with both bystanders and victims. His rationale is simple: "we should include the opinions of those bystanders who are likely to be burdened by intervention precisely because they are burdened bystanders: they are not (directly) responsible for the humanitarian crisis but might suffer in its resolution" (Pattison 2010: 144). If risk-imposition is what dictates input, as it does for both McMahan and Pattison, then *all* who are subjected to risk must be entitled to input.[19]

Once this claim is taken on board, however, a third group of individuals demands our attention. Suppose that an intervening entity were to consult with the foreign risk-bearers of an intervention before going into a state that was likely to retaliate with force against the intervening state. It may be that the relevant group of risk-bearers must include not just the humanitarian victims and not just bystanders in the target-state or nearby states but also a third group of individuals: citizens of the intervening state. Call these citizens *retaliation-risk-bearers* (or *indirect-risk-bearers*).

A key difference between these risk-bearers and those analyzed by Pattison is that the risks imposed upon retaliation-risk-bearers are not *directly* imposed by the intervening state, as they may be on victims and bystanders, but are *indirectly* imposed. In other words, the intervening entity itself would not be killing or harming retaliation-risk-bearers but would rather be provoking third parties to kill or harm retaliation-risk-bearers. But if indirect risk could be shown problematic in the way that direct risk is problematic, then the group of risk-bearers who would be entitled to participate in decisions of humanitarian intervention may include retaliation-risk-bearers in addition to the bystanders and intended beneficiaries who are subjected to risk.

In any case, the conclusion reached by Pattison and McMahan in the context of humanitarian intervention is seemingly pregnant with one further critical implication. If risk-bearers are entitled to participate in and potentially exercise veto power over humanitarian military authorization, why should they not be entitled to participate in *all* forms of military authorization? More to the point, if these individuals are entitled to participate in and potentially exercise veto power over military authorization, perhaps military authorization should *always* be reserved for public discharge. Much more needs to be said (and will be said in due course) about whether the collective participation of risk-bearers truly transforms, or ought to transform, a potentially private decision into a public one—and whether affected individuals *must ensure* this transformation. But we should note at least that the character of private military authorization would be dramatically altered by the participation of these risk-bearers.

Despite this suggestive line of reasoning, neither McMahan nor Pattison apply the risk-prevention argument beyond the context of humanitarian intervention, and neither considers possible connections to military privatization. Indeed, Pattison explicitly advances an alternative set of rationales for the public monopolization of military authorization.[20] Cécile Fabre (2012: 150–156), by contrast, considers an extension of the sort that I have in mind, even underscoring retaliation-risks, but she ultimately deems the extension unpersuasive. Fabre (2012: 151) imagines a scenario in which a minority group V has a just cause for war against A, and one individual from V authorizes that war:

> [H]is actions would in all likelihood lead A to inflict greater harms on V, for example, by way of reprisals. Even if the harm occasioned by this particular war would not violate the principle of proportionality, the objection would press, a lone individual does not have the authority, and thus lacks the right, to [go to war against A] ... at least without the consent of the individuals on whose behalf he acts—a consent which, as a lone individual, he is not in a position to secure.[21]

Fabre offers a number of objections to this version of the risk-prevention argument. I will rebut each over the course of the next two chapters.[22] But note that while the argument effectively identifies the indirect-risk-bearers that McMahan and Pattison neglect, it neglects the sort of direct-risk-bearers that they identify. Not only might individuals in V possess a right of participation in the decisions of private authorizing entities operating out of V's territory against State A (because of potential reprisals), but individuals from State A (the enemy) might also have a right of participation in these decisions. Individuals from State A are the direct-risk-bearers of any military authorization taken against them. Perhaps (at least some) potential victims of war in State A may be entitled to participate in the decisions of military authorization that are taken against their state just as the risk-bearers of a humanitarian intervention may be entitled to participate in decisions of military authorization that are taken against theirs. Lazar (forthcoming: 5) acknowledges as much, underscoring that "citizens of the hostile state" and "inhabitants of the broader region" raise "complex issues about whether all those affected by a decision should contribute to making that decision." But he balks at addressing these complex issues and neither endorses nor defends the inclusion of all affected individuals in decisions of military authorization.

3.3 All affected interests

The risk-prevention argument is a defense of precisely this inclusion. The principle underpinning my argument—and more specifically P_1—is "the all affected interests" principle. According to common renderings of this principle, individuals whose interests are likely to be affected by the decisions of a government ought to be included in the decisions of that government (e.g., as voters). The principle is typically deployed as a solution to the so-called boundary problem in democratic theory.[23] The boundary problem asks which individuals ought to be included in a *given demos*—who, in other words, ought to have a say in this democracy or that democracy. The rub is that the answer cannot be decided democratically. As Robert Goodin (2007: 43) writes, "[u]ntil we have an electorate, we cannot have an election," and it would be "incoherent to constitute the electorate through a vote among voters who would be entitled to vote only by virtue of the outcome of that very vote." A theory is needed that determines the make-up of electorates *before* the first election. The all affected interests principle stands as such a theory. It claims that individuals who are likely to be affected by the decisions of an electorate are the individuals who ought to be included in the decisions of that electorate.

In elucidating the potential scope of the all affected interests principle, Goodin draws our attention to an example, which will be frequently revisited in my defense of P_1.

> Imagine a German law that requires polluting factories there to build chimneys tall enough to ensure that their emissions fall to the ground only in

Scandinavia: legally, that law binds only manufacturers in Germany; but it clearly affects Scandinavians, and is indeed designed to do so.

(Goodin 2007: 50)

Excluding Scandinavians from a vote on this law would be problematic for proponents of the all affected interests principle, because the law would so thoroughly harm Scandinavians. As this example illustrates, the all affected interests principle may endorse expansive forms of inclusion. State decisions often harm people from afar and, for that reason, may entitle individuals beyond its borders to participate in its decisions.

But, as typically used, and as just presented, the all affected interests principle does not straightforwardly apply to the private authorization of military force. It explains *who* must be included in public decisions (all who are affected) without explaining why some decisions must be public in the first place—why affected individuals ought to remove certain decisions, like those of military authorization or environmental protection, from the private sector and reserve them for collective judgment. To connect the all affected interests principle, as presented, to a restriction on private decision-making, at least two further claims are needed.

The first claim is that *if* some public decisions require the inclusion of affected parties in the decision-making (the contention of the all affected interests principle, as typically framed), then surely some private decisions would also require the inclusion of affected parties in decision-making. The reason is that private decisions may exert effects that are no less profound than public decisions. Consider, for example, a German pharmaceutical company that, on its own accord, instituted a new industrial process for drug manufacturing that resulted in the release of harmful pollutants, which affected Scandinavians. If, according to the all affected interests principle, we insist that Scandinavians must have a say in laws that require German factories to spew harmful pollutants in their direction, then we must insist that Scandinavians have a say in private decisions that result in *these very pollution patterns*. The fact that a government severely affects Scandinavian interests in the first instance while a private company severely affects these interests in the second seems to be arbitrary if the Scandinavian entitlement to participation derives from disaffected interests. This simple point merely reflects the fact that, while governmental decisions often exert large-scale effects on individuals, private decisions may exert the identical effects.

The second claim is that meaningful participation in a private decision will sometimes demand that affected parties *collectively* exercise a *final say* over that decision. For some, the idea might seem dubious that Scandinavian input into the decisions of a German pharmaceutical company might demand that private decisions must be converted into public decisions (though I do not believe this to be dubious at all). So suppose instead that the German pharmaceutical company decides to release its harmful pollutants over German towns rather than Scandinavian towns. Because the pharmaceutical company, we may assume, has factories dotted all over Germany, all German residents will be affected by the decision. All will therefore have a strong interest in setting the level of pollution that is going to be released by the pharmaceutical company. It simply becomes difficult to imagine how German residents could adequately safeguard their interests unless the ultimate decision over pollution levels were publicly determined—unless the German government, as a representative of the German people, were to *regulate* how much pollution will be permitted by the pharmaceutical company (and other companies).

The difficulty, of course, even if this much is plausible, is in determining *which* private decisions ought to be transformed into (assuredly) public decisions on account of the all affected interests principle and which may (or must) remain private. A German pharmaceutical company that failed to consult with German citizens ahead of a small price-hike would probably be justified in doing so.[24] Yet, as we just saw, the company may not be justified in releasing harmful pollutants without broad public approval.[25]

To sidestep the difficult challenge of identifying *every* interest that may activate the all affected interests principle, I will restrict my analysis to (what I am dubbing) *fundamental interests* (hence the appellation of P_1 as the all affected fundamental interests principle). Fundamental interests are those interests that are so important as to be protected by rights. P_1 claims that private decisions that substantially affect *fundamental interests* must be withdrawn from the private sector. Whether decisions that affect other, non-fundamental interests must be similarly withdrawn is left aside. The authorization of war, I argue, affects *fundamental* interests. Thus, my conclusion that public actors ought to withdraw decisions that affect fundamental interests will be sufficient to show that public actors ought to withdraw decisions of military authorization from the private sector.

3.4 The all affected fundamental interests premise (P_1)

While these considerations should all lend *prima facie* plausibility to P_1, a principled defense of the premise is needed to firmly ground the risk-prevention argument. A close look at P_1 will reveal that it contains several potentially objectionable claims. To guard against any criticism, and to strip the premise of ambiguity, I will defend four separate sub-premises (P_1i, P_1ii, P_1iii, and P_1iv). The reader may skip ahead to Chapter 4 if these sub-premises are sufficiently persuasive without further defense.

P_1i *Fundamental Interests*: We have interests that are sufficiently weighty to be protected by rights (fundamental interests), one of which is our fundamental interest in physical security.

P_1ii *Participation*: When one course of action under consideration in a private decision will impose risk, above some threshold level, to my fundamental interests, then I am (or my representatives are), at a minimum, entitled to participate in that decision.

P₁*iii* *Oversight*: When I am (or my representatives are) entitled to participate in a private decision according to P₁*ii*, I am also duty-bound to support collective enforceable oversight of that decision. Such collective enforceable oversight constitutes a final say by those who are affected by the decision.

P₁*iv* *Public Discharge*: When *enough* individuals are duty-bound to exercise a final say over a private decision according to P₁*iii*, we should say that the decision ought to be withdrawn from the private sector and reserved for public discharge.

Taken together, these sub-premises form the claim of P₁: a decision ought to be withdrawn from the private sector and reserved for public discharge when one course of action under consideration imposes risk, above some threshold level, to the fundamental interests (e.g., physical security) of enough individuals.

3.4.1 Fundamental interests (P₁i)

The claim that we *do* have interests that are sufficiently weighty to be protected by rights is in keeping with an interest theory of rights. According to interest theories, a "necessary but insufficient" condition for the "holding of a right by a person X" is that the right "preserves one or more of X's interests" (Kramer 2000: 62). If an interest is ultimately to prove suitable for protection, further argumentation must show that it is sufficiently valuable to warrant protection. Joseph Raz (1986: 166) captures this further point in his suggestion that "X has a right if and only if" some "aspect of X's well being (his interest) is a sufficient reason for holding some other person to be under a duty."[26] The risk-prevention argument utilizes this Razian formulation of a right.

While some interests do not merit the protection of rights, one interest that certainly *does* merit the protection of rights, it seems, is our interest in physical security—and our interest in not being killed or disabled in those ways that make one a casualty of war. The reason why our interest in physical security must be protected by a right, on my account, and following Henry Shue (1996: 21), is that it is necessary for the enjoyment of all other rights: "[n]o rights other than a right to physical security can in fact be enjoyed if a right to physical security is not protected." For instance, without a right to physical security, rights that may protect our interests in receiving an education, forming close bonds with others, speaking and believing freely, and pursuing activities that lead to self-actualization and fulfillment are vulnerable. Thus, if we have a right to anything, we must have a right to physical security (the right to physical security is a *basic right* in that sense). And surely, in virtue of our human dignity and moral worth, we must have a right to *something*.[27] If this reasoning is sound, which I believe it is, then we may conclude that the interests of human beings in physical security are fundamental interests, because they warrant protection by rights (and indeed, they warrant protection by *basic* rights).[28]

3.4.2 Participation in private risk-imposing decisions (P₁ii)

In keeping with P_1i, when one course of action under consideration in a private decision will impose risk, above some threshold level, to my fundamental interests, it would seem that I am entitled to *something* vis-à-vis that decision. It cannot be, for instance, that the violation of my fundamental interest in physical security may proceed just as it would proceed if I lacked a fundamental interest in physical security. By the same token, it cannot be that the substantial imposition of risk, above some threshold level, to my fundamental interests in physical security may proceed just as it would proceed if I lacked a fundamental interest in physical security. The existence of a fundamental interest in physical security demands that decision-making be responsive, *in some sense*, to that fundamental interest. The difficult question is: in what sense should it be responsive? To what does a fundamental interest in physical security entitle an individual when a decision that may affect her fundamental interest in physical security is under consideration?

One response, which is also an objection to P_1ii, is that we are simply entitled to *not have our fundamental interests affected*. In other words, we possess a much more muscular entitlement than the modest demand of participation suggests. We ought to be free altogether from substantial risk-impositions. Consider Nozick's (1974: 74) example of a gunman exposing a passerby to risk by firing a partially loaded revolver. When the passerby is threatened, she does not wish to consult with the gunman. She wishes for the gunman to hold fire. Similarly, in the private pollution example above, it may seem that the German pharmaceutical company should simply *refrain from polluting* rather than include German residents in pollution-related decisions.

This objection to P_1ii is most compelling when the harm that violates our fundamental interests may be avoided with ease. Suppose, for instance, that the German pharmaceutical company faces a choice between releasing pollutants that will harm German residents and not releasing any pollutants that will harm German residents. When the costs of abstention are low and the benefits that are forfeited (due to abstention) are minimal, the company should undoubtedly choose to forego releasing harmful pollutants. In this case, a mere insistence that affected parties be included in decisions is too weak. We must go further and demand that parties not be affected.

But, as the drawbacks of avoiding these effects mount—when it becomes extremely problematic for the German pharmaceutical company to refrain from producing pollution—then the objection becomes less persuasive. Two particular kinds of drawbacks are worth noting. The first is *financial cost*. When the cost of foregoing pollution is steep, a difficult decision is needed about how much money will be spent in order to avoid the release of pollutants. The second kind of drawback is *competing risk*. If, when the pharmaceutical company cuts down on pollution, it produces less (or less effective) medication, a difficult decision is needed about the tradeoff between pollution and health care.[29] The decision *to* pollute will pose a risk to the fundamental interests of German residents, but the decision *not to* pollute will also pose a risk to the fundamental interests of German residents (if we suppose that our interest in health is a fundamental interest and that a disruption to the supply of medicine may violate that interest).

When a risk is imposed upon my fundamental interests, the entitlement that is generated in the face of high costs or competing risks cannot be a simple abstention from risk-imposing behavior. The results of abstention will be unpalatable if the costs of abstention are sufficiently high, for instance if the *full* renunciation of pollution would debilitate the German economy. Even worse, the results of abstention may be counter-productive and thus harmful to our fundamental interests if the competing risks of abstention are sufficiently high, for instance if the renunciation of *all* emissions would debilitate Germany's pharmaceutical industry. These considerations underscore the simple point that we are not always entitled to abstentions from risk-imposing behavior. At times, we are entitled to something less.

But why are we entitled to *participate* in decisions that affect our fundamental interests? The fact that German residents will be affected by the release of pollution from German pharmaceutical companies may just mean that the interests of affected parties ought to be reflected in pollution-related decisions. If so, then risk-imposing entities could simply factor the interests of affected parties into their decisions about pollution without the participation of these parties. The ultimate burden of defending P_1ii, it seems, is to explain why such private risk-imposing entities must not be permitted to make decisions on behalf of those whom they are likely to affect.

At least three reasons, I believe, may be given to justify the entitlement of participation that is promised by P_1ii.[30] First, risk-imposing entities like private companies are simply unlikely to promote the interests of those whom they affect on a purely voluntary basis. The individuals who will be affected by decisions will be more sensitive to any potential effects and more vigilant about steering decisions away from these effects.

A second more fundamental reason is that the individuals whose interests may be affected by a decision may disagree about the levels of risk that should be tolerated, all things considered. There may be "reasonable disagreement about what constitutes an acceptable risk, or an acceptable trade-off between risks and benefits" (McMahan 2010a: 53). Given this reasonable disagreement, consultation with individuals about what levels of risk will be tolerated at a given moment and for a given enterprise is demanded. If there were some known level of risk that would always be acceptable given the benefits and tradeoffs, the inclusion of potentially affected parties in the decision may be unnecessary. But there is no such common risk-tolerance that human beings possess. Instead, individuals must be given the opportunity to express their risk-tolerance on a case-by-case basis.

The third reason why individuals must be entitled to participate in decisions that affect their fundamental interests is that individuals have ill-defined risk-preferences. Participation ought to be guaranteed so that individuals can *develop* their risk-tolerance in a given context. This rationale neither presupposes that individuals possess the same levels of risk-tolerance, nor even that they possess a stable, well-defined level of risk-tolerance. Instead this third rationale gains its force from the recognition that individuals have notoriously unstable, even irrational, preferences regarding risk.[31] Individuals are poor at identifying how much risk is imposed upon them by a given course of action. We are poor at deciding whether that course of action imposes more or less risk than commonly performed tasks—like driving. Perhaps most troublingly, our preferences are easily manipulated by changes in the language with which risk-related questions are framed, even when that risk is made explicit.

Given our clumsy command of risk, individuals, I wish to suggest, must be entitled to participate in decisions that affect their fundamental interests in order to develop an informed view on that decision (which is then registered as a preference for or against the decision). Put differently, risk-imposing entities should not be permitted to make decisions on behalf of individuals who may be affected by those decisions, because this would prevent individuals from developing well-formed preferences in the first place. Why should we care about the development of well-formed preferences? Given the stakes of decisions that affect our fundamental interests, like pollution and war, a refusal to grant individuals an entitlement to develop well-formed preferences is a failure to adequately respect their fundamental interests. We ought to be afforded an opportunity, through participation, to develop well-formed preferences on decisions that may pose such profound implications for our welfare as war and pollution.

Note that none of these reasons precludes the possibility of electing political leaders to represent our fundamental interests. Within the confines of a representative democracy, I am still able to effectively safeguard my interests; political representatives may sort through the reasonable disagreement of risk-related decisions; and, through elections and political engagement, I may be able to develop well-formed preferences on complex risk-related decisions. My representatives, in turn, may exercise a final say over the sort of risk-imposing decisions under consideration. In short, the participation that is demanded of $P_1 ii$ is consistent with and possible within a representative democracy.

3.4.3 Obligations of oversight (P₁iii)

Having now demonstrated that individuals may be entitled to participate in private decisions that impose risks upon them ($P_1 ii$), let us now turn to the claim that these individuals have a *duty* to collectively exercise a final say over such decisions ($P_1 iii$). Would it be sufficient if the pharmaceutical company in our example above simply ascertained the risk-related preferences of *all* German citizens (say, in a nationwide poll) and relied upon these in decisions of risk-imposition? $P_1 iii$ insists that it would not.

The reason is that individuals who are entitled to participate in these high-stakes decisions of risk-imposition, as identified in $P_1 ii$, incur a duty to ensure that *others* who are subjected to similar risks are *properly* included; and fulfillment of this duty requires collective enforceable oversight of decisions. The same fundamental interest in physical security that generates my right to participate in private decisions of risk-imposition also generates a duty for me to help protect the fundamental interests of others who are, or ought to be, included in these decisions. To insist upon a right of participation with *no* corresponding duty to help guarantee the participation of others is to demonstrate inadequate respect for the fundamental interests that give birth to the right of participation in the first place. The duty to take *some* measure to ensure proper participation is possessed by all who are affected and owed to all who are affected. Individuals must aim to make certain that *their* decision does not assign excessive weight to the interests of some and inadequate weight to the interests of others.

What steps must individuals take? The most basic step, it would seem, is contribution to the establishment of a representative institution that could review the pharmaceutical company's decision and sign off on it (or not). If affected individuals owe each other anything (which they must), they surely owe each other some minimal institutional protection. And the most minimal of these protections (with any degree of efficacy) is *collective enforceable oversight*, which would enable the group to scrutinize decisions of the company and supersede those that contravene the procedural requirements of the all affected fundamental interests principle. An individual who was unwilling to contribute to collective enforceable oversight would be participating in a decision that unjustifiably treated her own interests as more valuable than the interests of other similarly affected individuals. But, of course, the collective entitlement to review decisions by the pharmaceutical company and override those that are improperly inclusive (or otherwise procedurally unjustifiable) *is* the entitlement to exert a final say (or trump) over such decisions, as demanded by $P_1 iii$.

Do executives at the pharmaceutical company have an obligation to submit to this collective enforceable oversight? They do, for precisely the reason given above. A refusal to do so would unjustifiably treat their own interests (fundamental or otherwise) as more valuable than the fundamental interests of all who are affected. Collective enforceable oversight is a duty owed by and to all affected individuals, and it is a duty whose fulfillment must not be impeded by private actors with the potential to improperly exclude affected parties.

3.4.4 *Withdrawal of decisions from the private sector ($P_1 iv$)*

As should be clear, once the need for collective enforceable oversight is established, the decision has been monopolized in the relevant sense. The group of affected individuals reserves final decision-making authority for itself. Decisions reached by the pharmaceutical company are therefore tentative in the sense that they are subject to the will of affected individuals.

However, imagine a scenario in which the pharmaceutical company faces a decision that affects the fundamental interests of only ten individuals in the area surrounding one of its factories. The decision poses a grave risk to the fundamental interests of these ten individuals, and all ten are therefore entitled to participate in the decision ($P_1 ii$). They fulfill their obligation to ensure the proper participation of one another (P_1iii) by establishing a council of oversight, and the pharmaceutical company fulfills its obligation to recognize the final decision-making authority of the council over pollution-related decisions. Would it be correct to say that this decision is now reserved for *public* discharge? I think that it would. But an opponent might insist that the collective of individuals is too small to be deemed "public." Supposing that we told the same story about a council that was constituted by the 80 million residents of Germany who, together, issued binding directives regarding appropriate pollution levels for the pharmaceutical company, *then* (the objection would press) the decision is "public." If this objection is correct, then the requirement of public decision-making that is the payoff of P_1 may be called into question.

But notice how this objection is simply definitional; if true, it would not undermine the chapter's central argument that individuals whose fundamental interests are subjected to risk (above some threshold) have a right and a duty to collectively exercise a final say over such risk-imposition. Nevertheless, because war necessarily imposes risk on *a lot* of individuals (as we will see in the next chapter), the risk-prevention argument is amenable to the following stipulation: when *enough* individuals possess the right and duty to collectively exercise a final say over a decision of the sort identified by P_1ii and P_1iii, then it will be appropriate to say that the decision ought to be publicly discharged (and withdrawn from the private sector). Some will take this definitional concession to be superfluous and others necessary. But it captures the intuition that public decisions are collective *and* involve a large number of individuals. The goal of the next chapter is to provide a theory of monopolization in the realm of war—which affects many individuals—and so this concession should allay any concern that the risk-prevention fails to deliver a requirement of *public* monopolization.

3.5 Conclusion

Two final points are worth making before concluding this chapter. First, my defense of P_1 has focused on the most skeletal of public monopolies—one of collective enforceable oversight. But, of course, my defense is consistent with more muscular monopolization: for instance, a monopoly in which the collective itself holds votes on risk-imposing decisions, organizes forums for conversation and deliberation, sets rules related to campaign finance, and so on. I have focused on a single, publicly minded pharmaceutical company, which aimed to ensure that its decisions were in keeping with the preferences of all who might be affected, in order to show that, even in these optimal circumstances, the collective must retain final decision-making authority. Pharmaceutical companies, of course, do not *actually* behave like the one in this chapter. Most tend to pursue their own interests, not the interests of all who are affected by their decisions (perhaps, to be maximally charitable, because they are already subject to more robust regulation than has been defended in this chapter). Far more expansive monopolization is therefore likely to be needed than the minimal monopolization I identify.

Second, the duty of collective enforceable oversight that I have defended is potentially capable of justifying a rudimentary state. Individuals living in some swath of territory face the *risk* that any number of entities will make risk-imposing decisions (the risk of a risk). Thus, it would seem that individuals in this territory may be duty-bound to one another to establish institutions that monitor whether *any* company (or other set of individuals) are making decisions that affect their fundamental interests. Perhaps all who may be affected by potential decisions have a duty to erect preventive institutions that review and potentially override decisions that impose unjustifiable risks. Such an institution begins to look a lot like a state regulating behavior within some geographical territory. But this, needless to say, is an implication that the risk-prevention argument would gladly embrace. A full exploration of this implication, however, is much beyond the scope of this book.

To summarize, I began this chapter by considering five prominent defenses of public monopolization: the Thomist view that individuals must defer to superior legal institutions, the Kantian proposition that they must resolve reasonable disagreement through fair deliberation, the consequentialist insistence that the dangers of non-monopolization are too great, the democratic worry that non-monopolization would threaten the instrumental or intrinsic value of self-governance, and the egalitarian (or sufficientarian) demand that public monopolization is a requirement of distributive justice. None of these arguments succeeded in establishing a requirement of public military monopolization. But each underscored the extensive array of questions that a theory of public monopolization and legitimate authority must answer if it is to be persuasive.

I then sketched a two-premise defense of the risk-prevention argument for public monopolization, which built upon the work of McMahan, Pattison, Fabre, and Lazar. After sketching these two premises, I devoted the remainder of the chapter to a justification of the first premise: the all affected fundamental interests premise (P_1). This premise borrowed a principle that is familiar to theorists of the boundary problem—the all affected interests principle—and applied it to debates on privatization. For the sake of argumentative ease, I focused specifically on *fundamental interests* (those interests that are protected by rights), and I presented four claims. First, we possess fundamental interests in physical security (P_1i). Second, these fundamental interests entitle individuals to participate in private decisions that pose considerable risks to these interests (P_1ii). Third, fundamental interests entail duties to help ensure collective enforceable oversight of risk-imposing decisions (P_1iii). Fourth, the collective oversight is public when enough individuals must help ensure collective enforceable oversight (P_1iv).

Notes

1 This example, on first glance, will undoubtedly raise a number of questions and concerns. I cannot address these questions and concerns at this stage but will do so in due course.

2 By "withdrawn from the private sector and reserved for public discharge," I mean that all affected individuals ought to *ensure* that *they* collectively exercise a final say over

decisions of military authorization by preventing private actors from exercising this final say.

3 Each of these five arguments presumes (or defends the claim) that public monopolizing entities may also themselves authorize military force. Pacifist endorsements of public monopolization will be left aside for now, largely because my own position aims to fill a gap in the ethics of war tradition. But, in Section 4.5, once the risk-prevention argument is complete, I will consider how my position compares with a pacifist position on public monopolization (and authorization). There, I challenge the pacifist rejection of public military authorization that accompanies their embrace of public military monopolization.

4 This is quoted in Caney (2005: 206).

5 Pattison (2008: 150) is careful to note that his arguments do not seek to show that public military authorization is a necessary condition for the permissibility of a given war. Rather, public military authorization "is an important … factor in the overall justice of war."

6 Reitberger (2013: 85) considers (and dismisses) a related argument. The argument is that even if democratic legitimate authority may not promote peace, it at least "increases the likelihood that wars will be just." This argument also suffers from empirical uncertainty.

7 Pattison (2014: 169) points to Avant (2006: 520) and Isenberg (2009: 12) for a similar remark. See also Fabre (2012: 149) and Coates (1997: 131–134).

8 One worry, which I will analyze at great length, is that private actors may drag their states into war. In that sense, these actors may make it difficult for citizens ultimately to choose a course of action besides war. However, values of self-government and individual autonomy, it would seem, are still realized in those circumstances in which citizens democratically authorize wars since individuals are still making a choice.

9 Of course, the regrettable loss may be that citizens who do not monopolize military authorization will be engaged in one less collective enterprise, namely the collective enterprise of preventing private actors from authorizing military force. But this does not seem to be the worry at stake.

10 For a powerful critique of the value of self-government, see Valdman (2010).

11 Pattison (2010: 129–151) is a revised version of Pattison (2007).

12 Pattison (2014: 218), for example, has argued that, insofar as private authorization of military force is permitted in a system, the poor are likely to suffer from under-provision, and the rich are likely to enjoy greater protection.

13 It is true that individuals in a given polity may receive varying amounts of protection. But national defense today is *largely* perceived by advanced democracies as holistic. Any breach is thought to be a threat to all.

14 Pattison refers to "burdens," while McMahan refers to "risks." Since there is a chance (albeit slim) that a humanitarian intervention will not *actually* harm intended beneficiaries, I prefer a language of "risks" to "burdens." But it is clear that Pattison is concerned with risks just like McMahan. Pattison (2010: 143) worries, for example, that the intended beneficiaries "*might have to suffer* civilian and military casualties, damage to vital infrastructure, increased levels of insecurity, and other costs associated with being in a war zone" (my emphasis added).

15 See also McMahan (2005: 13) for the conclusion that the justifiability of humanitarian intervention depends upon the consent of beneficiaries, but his argumentation for this conclusion is provided only in later work (McMahan 2010a: 142–144).

16 See Altman and Wellman (2008: 242–245) for the opposing view that authorizing entities need not obtain consent from the beneficiaries of humanitarian intervention.

17 Tesón (2003: 105–107) rejects the need for consultation with the first two groups, endorsing only consultation with the third.

18 Of course, interveners may intend to benefit victims of a humanitarian catastrophe *and* others. But the analysis remains unchanged.

19 McMahan's failure to demand that bystanders be included in consultation prior to intervention is perplexing given his argument elsewhere that imposing lethal risks on bystanders without their consent is *worse* than imposing lethal risks on those who stand to benefit from risk-imposition without their consent (McMahan 2010a: 62–69, 2010b). On his account, "beneficiaries rather than bystanders ought, if possible, to bear the unavoidable costs of their own defense" (McMahan 2010b: 361). Thus, it seems that McMahan has particular reason to insist that bystanders be included in decisions of military authorization: since non-consensual risk-imposition upon bystanders is *ceteris paribus* worse than non-consensual risk-imposition upon beneficiaries, on his account, and since beneficiaries must be included in decisions of authorization, then surely bystanders must also be included in decisions of authorization.

20 See the final three arguments in Section 3.1 above for several of Pattison's arguments for legitimate authority. To my knowledge, McMahan does not offer a detailed treatment of legitimate authority in the broader context of war.

21 The view on offer by Fabre should be distinguished from a related view. The related view claims that private individuals must not authorize military force against State A because to do so would be to undermine the right of self-government of those on behalf of whom war is authorized. Pattison (2010: 137–139) presents this argument in the context of humanitarian intervention and war more generally (Pattison 2014: 80); Finlay (2010) presents it in the context of political liberation. To distinguish between the risk-imposition argument and these views, suppose that the people on behalf of whom an authorizing entity acts will *not* be risked by an intervention. They will remain perfectly safe. Still, according to Pattison and Finlay, the authorizing entities commit a wrong.

22 Keep in mind that Fabre targets much of her criticism at the perceived requirement of consent in the risk-prevention argument. But as we have seen, consent, strictly speaking, may not be necessary for the argument's success. McMahan and Pattison persuasively insist upon something short of consent—namely democratic representation—in the context of humanitarian intervention.

23 For key works on the boundary problem, see Dahl (1979: 97–133), Whelan (1983: 13–47), Arrhenius (2005: 14–29), Goodin (2007), Miller (2009), List and Koenig-Archibugi (2010). The problem has also been dubbed the problem of "constituting the demos" and "the problem of inclusion."

24 For this example used in the context of supermarket price-hikes, see Miller (2009: 217).

25 Notice, however, that the identical difficulty confronts the all affected interests principle when applied to the boundary problem. We might think that the German decision to direct pollution at Scandinavian countries would require the inclusion of Scandinavians in the demos for pollution-related decisions but not in the German decision to raise tariffs by 1 percent. Such a decision would affect the interests of Scandinavians—at least those who trade with Germany—but presumably not in such a way that would entitle them to participation.

26 Raz also points out that X must be able to possess rights.

27 Note that, on my understanding, fundamental interests are not always protected by basic rights (among which Shue includes the rights to physical security, subsistence, and liberty). They may be protected by other non-basic rights. Note also that Shue's argument for basic rights is not dependent upon an interest theory of rights.

28 Typically, interest theories of rights, like the one embraced here, are contrasted with will theories. For a powerful defense of will theories, see Steiner (2000: 233–306), who relies to an extent on Hart (1982: 162–193). For helpful comparisons of interest and will theories, see Wenar (2005: 2011) and Kramer and Steiner (2007). Unfortunately, a defense of interest theories is much beyond the scope of this book. But even a committed will theorist ought to be persuaded by my position. Will theorists neither

reject the possibility of rights protecting interests, nor do they dismiss the deep value of human interests. They only object to the notion that rights *necessarily* protect interests. The will theorist is therefore free to acknowledge that human beings have a profound interest in physical security—which may be protected by a right, just not in virtue of that fact—and that, in the end, such a profound interest may demand the withdrawal of decisions from the private sector that affect this interest (if, of course, P_1ii, P_1iii, and P_1iv are sound). No mention of rights is required. To put this point differently, the term "rights" has *only* been used in my argument to describe a set of deeply valuable interests. In foregoing this term, the interests do not become any less valuable. They simply go by a different name.

29 It is worth noting here that the decision to authorize war is particularly vulnerable to the second kind of drawback. If war is *not* authorized by a group under threat, the risks to that group may be high. And if war *is* authorized by a group under threat, the risks to that group may be high. This consideration becomes relevant in the next chapter.

30 These reasons, we should note, apply both to public and private decisions. I refer only to private decisions here since they are the decisions with which P_1ii is concerned. But we might apply each of these reasons to public decisions—for instance, to argue that the German decision to export pollution to Scandinavia ought to include Scandinavians in their decisions.

31 See, for instance, Kahneman (2003) for a *locus classicus* of this enormous behavioral economics literature.

References

Altman, Andrew and Christopher Heath Wellman (2008). "From Humanitarian Intervention to Assassination: Human Rights and Political Violence," *Ethics*, Vol. 118, No. 2, pp. 228–257.

Aquinas, Thomas (2003 [1266–73]). "Excerpts from *Summa Theologiae*," in William P. Baumgarth and Richard J. Regan (eds), *On Law, Morality, and Politics*, second edition (Indianapolis: Hackett Publishing Company).

Arrhenius, Gustaf (2005). "The Boundary Problem in Democratic Theory," in Folke Tersman (ed), *Democracy Unbound: Basic Explorations I* (Stockholm: Filosofiska Institutionen, Stockolms Universitet), pp. 14–29.

Avant, Deborah (2006). "The Implications of Marketized Security for IR Theory: The Democratic Peace, Late State Building, and the Nature and Frequency of Conflict," *Perspectives on Politics*, Vol. 4, No. 3, pp. 507–528.

Caney, Simon (2005). *Justice Beyond Borders: A Global Political Theory* (Oxford: Oxford University Press).

Coates, A.J. (1997). *The Ethics of War* (Manchester: Manchester University Press).

Dahl, Robert (1979). "Procedural Democracy," in P. Laslett and J.S. Fishkin (eds), *Philosophy, Politics & Society* (Oxford: Blackwell), pp. 97–133.

Fabre, Cécile (2012). *Cosmopolitan War* (Oxford: Oxford University Press).

Finlay, Christopher J. (2010). "Legitimacy and Non-State Political Violence," *The Journal of Political Philosophy*, Vol. 18, No. 3, pp. 287–312.

Goodin, Robert (2007). "Enfranchising All Affected Interests, and Its Alternatives," *Philosophy and Public Affairs*, Vol. 35, No. 1, pp. 40–68.

Hart, H.L.A. (1982). *Essays on Bentham* (Oxford: Oxford University Press).

Held, Virginia (2005). "Legitimate Authority in Non-State Groups Using Violence," *Journal of Social Philosophy*, Vol. 36, No. 2, pp. 175–193.

Isenberg, David (2009). *Shadow Force: Private Security Contractors in Iraq* (Westport: Praeger Security International).

Kahneman, Daniel (2003). "Maps of Bounded Rationality: Psychology for Behavioral Economics," *The American Economic Review*, Vol. 93, No. 5, pp. 1449–1475.

Kramer, Matthew H. (2000). "Rights Without Trimmings," in Matthew Kramer, N.E. Simmons, and Hillel Steiner (eds), *A Debate Over Rights: Philosophical Enquiries* (Oxford: Oxford University Press), pp. 7–111.

Kramer, Matthew H. (2007) and Hillel Steiner, "Theories of Rights: Is There a Third Way?," *Oxford Journal of Legal Studies*, Vol. 27, No. 2, pp. 281–310.

Lazar, Seth (forthcoming). "Authorisation and the Morality of War," *Australasian Journal of Philosophy*, pp. 1–19.

List, Christian and Mathias Koenig-Archibugi (2010). "Can There Be a Global Demos? An Agency-Based Approach," *Philosophy and Public Affairs*, Vol. 38, No. 1, pp. 76–110.

McMahan, Jeff (2005). "Just Cause for War," *Ethics and International Affairs*, Vol. 19, No. 3, pp. 1–21.

McMahan, Jeff (2010a). "Humanitarian Intervention, Consent, and Proportionality," in N. Ann Davis, Richard Keshen, and Jeff McMahan (eds), *Ethics and Humanity: Themes from the Philosophy of Jonathan Glover* (Oxford: Oxford University Press), pp. 44–72.

McMahan, Jeff (2010b). "The Just Distribution of Harm Between Combatants and Non-combatants," *Philosophy and Public Affairs*, Vol. 38, No. 4, pp. 342–379.

McPherson, Lionel (2007). "Is Terrorism Distinctively Wrong?," *Ethics*, Vol. 117, No. 3, pp. 524–546.

Miller, David (2009). "Democracy's Domain," *Philosophy and Public Affairs*, Vol. 37, No. 3, pp. 201–228.

Nozick, Robert (1974). *Anarchy, State, and Utopia* (New York: Basic Books).

Pattison, James (2007). "Representativeness and Humanitarian Intervention," *Journal of Social Philosophy*, Vol. 38, No. 4, pp. 569–587.

Pattison, James (2008). "Just War Theory and the Privatization of Military Force," *Ethics and International Affairs*, Vol. 22, No. 2, pp. 143–162.

Pattison, James (2010). *Humanitarian Intervention and the Responsibility to Protect* (Oxford: Oxford University Press).

Pattison, James (2014). *The Morality of Private War* (Oxford: Oxford University Press).

Raz, Joseph (1986). *The Morality of Freedom* (Oxford: Oxford University Press).

Reitberger, Magnus (2013). "License to Kill: Is Legitimate Authority a Requirement for Just War?," *International Theory*, Vol. 5, No. 1, pp. 64–93.

Rothbard, Murray N. (1973). *For a New Liberty: The Libertarian Manifesto* (New York: Macmillan).

Schwenkenbecher, Anne (2013). "Rethinking Legitimate Authority," in Fritz Allhoff, Nicholas G. Evans, and Adam Henschke (eds), *Routledge Handbook of Ethics and War: Just War Theory in the Twenty-First Century* (Oxford: Routledge), pp. 161–170.

Shue, Henry (1996). *Basic Rights: Subsistence, Affluence, and U.S. Foreign Policy*, second edition (Princeton: Princeton University Press).

Steiner, Hillel (2000). "Working Rights," in Matthew Kramer, N.E. Simmons, and Hillel Steiner (eds), *A Debate Over Rights: Philosophical Enquiries* (Oxford: Oxford University Press), pp. 233–306.

Stilz, Anna (2014). "Authority, Self-Determination, and Community in *Cosmopolitan War*," *Legal Theory*, Vol. 33, No. 3, pp. 309–335.

Tesón, Fernando (2003). "The Liberal Case for Humanitarian Intervention," in J.L. Holzgrefe and Robert O. Keohane (eds), *Humanitarian Intervention: Ethical, Legal, and Political Dilemmas* (Cambridge: Cambridge University Press), pp. 93–129.

Thompson, Janna (2005). "Terrorism, Morality, and Right Authority," in Georg Meggle (ed), *Ethics of Terrorism & Counter-Terrorism* (Frankfurt: Ontos Verlag), pp. 151–160.

Valdman, Mikhail (2010). "Outsourcing Self-Government," *Ethics*, Vol. 120, No. 4, pp. 761–790.

Valls, Andrew (2000). "Can Terrorism Be Justified?," in Andrew Valls (ed), *Ethics in International Affairs: Theories and Cases* (Lanham: Rowman & Littlefield), pp. 65–79.

Wenar, Leif (2005). "The Nature of Rights," *Philosophy and Public Affairs*, Vol. 33, No. 3, pp. 223–252.

Wenar, Leif (2015). "Rights," in Edward N. Zalta (ed), *The Stanford Encyclopedia of Philosophy* (Fall 2015 Edition). Available at http://plato.stanford.edu/entries/rights/ (accessed December 28, 2015).

Whelan, F.G. (1983). "Democratic Theory and the Boundary Problem," in J.R. Pennock and J.W. Chapman (eds), *Liberal Democracy* (New York: New York University Press), pp. 13–47.

4   The risk-imposition of war

With a defense of the all affected fundamental interests premise (P_1) complete, let us now turn to the second premise of the risk-prevention argument: the *risk-imposition of war premise* (P_2). According to P_2, the private decision to authorize military force *does* impose risk, above an acceptable threshold, to the fundamental interests of individuals such that the decision ought to be withdrawn from the private sector and reserved for public discharge. The successful completion of the risk-prevention argument will solidify our rejections of A_1 and A_2 by affirming the public responsibility of military *monopolization*.

The argument does not, however, affirm the public permission of military *authorization*. As certain pacifist theories of war require, public actors may possess a duty to prevent private actors from authorizing war but lack a duty to authorize war themselves. I will therefore supplement the risk-prevention argument with what I have dubbed the *compensation argument* to defend the public permission of military authorization. According to the compensation argument, public monopolizing entities must authorize military force when those who have been disarmed are threatened, because these entities would otherwise violate a duty of care that is owed to those who have been disarmed. This argument builds upon a position developed by Robert Nozick (1974: 110–113). If this position is sound, then A_0, the non-provision of military force, must be rejected.

My analysis in this chapter proceeds as follows. I begin by showing that the private decision to authorize military force imposes risks to the fundamental interests of *many* individuals; this analysis contributes to P_2 by providing the arguments that are necessary to identify the individuals whose interests *count* when tallying the overall effects. The first section specifies a set of direct-risk-bearers whose fundamental interests are affected by private decisions of military authorization and argues that some possess an entitlement to participate in these private decisions.[1] The second and third sections together identify a set of indirect-risk-bearers whose fundamental interests are affected by private decisions of military authorization. I argue that, like direct-risk-bearers, some indirect-risk-bearers possess an entitlement to participate in these private decisions. Specifically, I defend two claims: the *heightened probability claim* and the *moral relevance claim*. The former is defended in the second section and the latter in the third (remember that the *affiliates* of private authorizing entities are those who may become targets of retaliation on account of their shared group membership with, or proximity to, private authorizing entities):

(1) *Heightened Probability*: The private decision to authorize military force heightens the probability (and hence risk) that the fundamental interests of the authorizing group's affiliates will be contravened (via the mechanism of indirect risk-imposition).
(2) *Moral Relevance*: The fact that the private decision will not *directly* impose risk upon the affiliates but will only *indirectly* impose such risk (by causing a third party to directly impose risk upon them) is not sufficient to eliminate the affiliates; entitlement of participation.

If these claims are successfully defended, I will be in a position to confirm a close approximation of P_2: the private decision to authorize military force imposes direct and indirect risks to the fundamental interests of a lot of individuals, and it does so in a sense that is relevant for application of the all affected fundamental interests principle (P_1).

The fourth section then argues for a final proposition that converts this approximation of P_2 into a full-fledged endorsement. The final proposition is that, given the nature of war, the magnitude of potential contraventions to the fundamental interests of direct-risk-bearers and indirect-risk-bearers is sufficiently high (and for enough individuals) to place the heightened probability of contravention beyond a threshold of risk that would be acceptable for private military authorization to be justifiable. If this is so, then P_2 must be endorsed, and the decision to authorize military force ought to be withdrawn from the private sector and reserved for public discharge.

Once the risk-prevention argument is complete, the final three sections tie up several loose ends. The fifth section presents the compensation argument against A_0 for public military authorization. The sixth section considers emergency circumstances when the collective decision-making demanded of the risk-prevention and compensation arguments is not possible. The seventh section rebuts several objections to the position that I have defended. The successful rebuttal of these objections will allow us to move from Part I (and its analysis of A_0–A_2) to Part II (and its analysis of A_3–A_5) with an account of why public actors ought to monopolize and, when necessary, authorize military force.

4.1 Direct-risk-bearers

Which direct-risk-bearers may be entitled to participate in decisions of military authorization? We should recall, as highlighted at the start of Chapter 3, that one set of direct-risk-bearers who should certainly *not* be entitled to participate in decisions of military authorization are those direct-risk-bearers who have forfeited their immunity from military attack—which is to say, from substantial risk-imposition. If a general from State A leads an unprovoked invasion into State B, and State B responds with military force against State A in self-defense,

State B may subject the general from State A to substantial risks (by, for instance, trying to kill him). But the general should not be entitled to participate in the decisions of State B. This is because he has forfeited his immunity from substantial risk-imposition by attacking State B, and such immunity is necessary for any inclusion in the state's decisions. Without immunity from substantial risk-imposition, one has no grounds upon which to invoke the all affected fundamental interests principle and thus no grounds upon which to demand the recognition of rights (e.g., to physical security) that generate the entitlement to inclusion.

The first task then for anyone seeking to specify which direct-risk-bearers may be entitled to participate in decisions of military authorization is to explain which individuals retain their immunity from attack in war. Not all such individuals ultimately merit an entitlement of participation, as I will argue in a moment. But the retention of immunity is a necessary condition for the entitlement of participation.

Our views about who retains their immunity from attack will depend upon deeper views regarding (*i*) the moral equality of soldiers and (*ii*) non-combatant immunity—whether, to be more specific, we endorse a traditionalist just war account pushed by theorists like Michael Walzer (2006b: 34–50, 138–159; 2006c) or we instead endorse a revisionist account pushed by theorists like Jeff McMahan (2004; 2006a; 2006b; 2008; 2009). According to the traditionalist account, soldiers are typically liable to attack in war, while civilians are immune. According to the revisionist account, some soldiers may in fact be immune from attack, while some civilians may be liable. Traditionalists, in short, embrace principles of moral equality and strict non-combatant immunity that revisionists reject.[2]

To circumvent this debate—one that reflects perhaps the most intractable gulf among political theorists now writing about war—I want to suggest modestly that, for both traditionalists and revisionists, a number of direct-risk-bearers, including *both* civilians and soldiers, will retain their immunity from substantial risk-imposition in war. Whether these direct-risk-bearers merit a further entitlement to participate in decisions of military authorization will be addressed after the topic of immunity has been treated.

4.1.1 Civilian immunity from risk-imposition

The vast majority of civilians, according to both traditionalists and revisionists, ought to be immune from substantial risk-imposition in war. For traditionalists, who embrace a strict principle of non-combatant immunity, *all* civilians retain their immunity. For revisionists, all civilians except those who contribute to the creation of unjust threats retain their immunity.

Civilians, we should note, not only become direct-risk-bearers in military confrontations where the principles of *jus in bello* are violated. They become direct-risk-bearers in almost every war, even wars fought in strict accordance with international law and morality. Wars are fought in physical spaces, and because the spaces that civilians occupy are never fully sheltered from the spaces that wars occupy, civilians are always threatened, and almost always killed or injured in every war, despite the level of due-care that is exercised.[3] This is how wars go and part of what makes them terrible: they pull civilians who inhabit spaces off the battlefield into spaces where battle occurs.

This point should also remind us that some civilians who become direct-risk-bearers in war (and perhaps even some military personnel) are neither on the just side nor the unjust side but are simply *near* one of the sides. For both traditional-ists and revisionists, these physically proximate individuals would retain their immunity from substantial risk-imposition.

4.1.2 Military immunity from risk-imposition

The set of military personnel who retain their immunity from substantial risk-imposition in war is similarly wide according to both traditionalists and revision-ists. But a slightly more nuanced analysis is required given the complexity of disagreement between traditionalists and revisionists over the moral equality of soldiers. On a simplified understanding of the traditionalist position, *no* military personnel retain their immunity from substantial risk-imposition in war. Revi-sionists, by contrast, insist that *many* military personnel retain their immunity, namely all who serve on the just side.

But I believe this simplified version is an incorrect characterization of the tra-ditionalist position. Traditionalists do not maintain that all military personnel forfeit their immunity from attack vis-à-vis the entities that authorize military force. Consider the private authorization of military force by Al-Qaeda on Sep-tember 11 against the Pentagon, where military personnel were housed.[4] Though the soldiers of authorizing entities may not commit wrongs when targeting other soldiers, according to traditionalists, the *authorizing entity* certainly commits a wrong against soldiers when it initiates war against them without just cause. Walzer (2006b: 28, 31), for instance, writes that those who aggress, like Al-Qaeda, "are responsible for the pain and death that follow from their decisions, or at least for the pain and death of all those who do not choose war as a per-sonal enterprise"—choosing, he continues, "effectively disappears as soon as fighting becomes a legal obligation or a patriotic duty." In other words, many military personnel, particularly those who have been conscripted or who have joined the military out of patriotic obligation, have done nothing to forfeit their immunity from substantial risk-imposition imposed by the authorizing entity (nor have the civilian leaders of these soldiers done anything to jeopardize that immunity) according to traditionalists. For traditionalists then, these military personnel would not relinquish any potential entitlement to participate in deci-sions of military authorization on account of having forfeited their immunity from risk-imposition vis-à-vis the authorizing entities that make these decisions.

What do revisionists say about such military personnel? As we have seen, for revisionists, all such military personnel on the just side will retain their immu-nity from substantial risk-imposition in war. But, even for revisionists, those military personnel who are *involuntarily* on the unjust side (perhaps in the way that Walzer identified) also retain their immunity from substantial risk-imposition in war. The group of military personnel who thus retain their freedom from substantial risk-imposition in war will include many on the just and unjust sides who have not chosen, in the full sense of the term, to fight in the military.

4.1.3 From civilian and military immunity to participation

Are all (or any) of the civilians and military personnel who retain their immunity from substantial risk-imposition in war entitled to participate in decisions of private military authorization that are directed against them? After all, these decisions nevertheless involve the use of military force against *their* compatriots and geographical neighbors. When Al-Qaeda was considering the authorization of military force against the US, we must ask, were they really required to consult with direct-risk-bearers in the *US*?

Though perhaps counter-intuitive, the answer, I think, is yes. Assuming risk-imposition of a certain severity, direct-risk-bearers who have done nothing to forfeit their immunity from substantial risk-imposition ought to be included in such private decisions of military authorization *with one caveat*. The one caveat is that those direct-risk-bearers who are able to meaningfully voice their input with the group against whom private military force is authorized should not be included in such private decisions—despite the fact that these individuals have done nothing to forfeit their immunity from substantial risk-impositions.

Suppose, for instance, that a powerful group of private actors is set to author-ize massive military force against an unjust threat, and the unjust threat is a democratic state. The citizens in that state, we may presume, voted in elections and participated in the democratic process that resulted in the formation of the unjust threat, though they themselves were opposed to its creation. Should these individuals, who have had a meaningful say in the decision of their government, then get a second say in private decisions directed against their government?

The answer, I think, is no. But we are assuming *meaningful* democratic parti-cipation. One way that a citizen might lack a meaningful chance to protect her fundamental interests is if she is shut out (say, by an autocratic government) of the decision that prompts private military authorization. A second way is if there is no decision that prompts private military authorization—if the private actor authorizes military force without specific provocation. When individuals lack a fair chance to protect their fundamental interests through political participation at home, then these individuals, it seems, *should* get a say in decisions that are directed against them by private actors, decisions in which they *are* able to protect their fundamental interests.[5]

Any reasons that might prompt us to think otherwise look to be implausible. Neither physical location nor national association should be sufficient to elim-inate the individual entitlement to participate in risk-imposing decisions. As we saw in Section 3.3, when considering the all affected interests principle, Scandi-navians were plausibly entitled to express a say in pollution-related decisions made by the German government when these decisions affected their interests. Indeed, because pollution-related decisions affect the interests of far-flung individuals, the demos for such decisions may be required to significantly expand. The same should be said of risks in the context of military authorization. If the all affected fundamental interests principle is taken seriously, then we must conclude that *all whose fundamental interests are affected* by a decision of military authorization—bar those who have forfeited their immunity from substantial risk-imposition or who have enjoyed a meaningful say elsewhere—be included in that decision regardless of where the individuals reside.

Similarly, whether the decision of military authorization is public or private should be incapable of preventing application of the all affected fundamental interests principle. As we have seen, the all affected fundamental interests principle applies to *all* decisions that affect our fundamental interests. The fact that an entity is private rather than public is insufficient to absolve such an entity from demands that are generated by the all affected fundamental interests principle. The inclusion of direct-risk-bearers who are immune from risk-imposition in private decisions of military authorization should depend upon the risks that are imposed on their fundamental interests (with the one caveat underscored above). It should *not* depend upon the nationality or location of the risk-bearers, nor upon whether the risk-imposing entity happens to be a private rather than public agent.

The final point to highlight regarding the direct-imposition of risk is that the number of individuals who are entitled to participate in private decisions of military authorization may balloon when targets of private military authorization are unknown or uncertain. I have assumed that the direct-risk-bearers in the September 11 attacks on the US were limited to US residents since it was these individuals who were eventually attacked. But, whenever members of a group like Al-Qaeda choose to authorize military force, the potential targets of attack unfortunately abound. One might argue then that, in contemporary international relations, the direct-risk-bearers of a *given* private decision of military authorization ought to include individuals located around the world. As a United Nations High Level Panel (2004: 14) put the point: "[t]oday, more than ever before, threats are interrelated and a threat to one is a threat to all." This language is somewhat hyperbolic. But it underscores the sense in which the risk-prevention argument may call for the entitlement of numerous individuals to participate in decisions of military authorization.

Of course, such a proposition may seem radical. But if we pause for a moment to reflect upon the payoff of this proposition, we will see that it is actually somewhat modest (and is indeed a common refrain in international political discourse). Because wars are unpredictable and because they have the potential to spread and spiral out of control, the direct-risk-bearers in a given decision of military authorization may be so wide that the decision should not only be withdrawn from the private sector but should actually be transformed into a *public international decision*.[6] If so, the risk-prevention argument would necessitate strengthened international institutions like the UN. My conclusion, however, would only be radical insofar as current UN ambitions, which aim to "take effective *collective* measures for the prevention and removal of threats to the peace," are radical (United Nations 1945: art. 1.1). All that may be required by a rigorous application of the all affected fundamental interests principle is further international cooperation to guarantee the input of risk-bearers.

But we are getting ahead of ourselves. Let us consider indirect-risk-bearers in war before reflecting any further on the global implications of the risk-prevention argument. These implications will be considered after the risk-prevention argument has been finalized.

4.2 Indirect-risk-bearers: heightened probability

Like direct-risk-bearers, indirect-risk-bearers, on my account, are similarly entitled to participate in decisions of military authorization. Private authorizing entities impose indirect (or retaliation) risk on individuals when these authorizing entities provoke the deployment of other militaries that in turn directly impose risk upon them. Two claims—the heightened probability claim and the moral relevance claim—will help solidify our commitment to the inclusion of indirect-risk-bearers in private decisions of military authorization.

4.2.1 Heightened probability in the current international system

According to the heightened probability claim, the private decision to authorize military force heightens the probability that the fundamental interests of the authorizing group's affiliates will be contravened via the mechanism of indirect-risk-imposition. This is true, I wish to suggest, for both the current state-based international system *and* stateless international systems.

In the current international system, the mechanism for the heightened probability claim is this:

> *Current International System*: When the coordination required of private military authorization is undertaken by group G in some physical location L against the residents s_1-s_n of state S, the state T in which L is located becomes the target of increased probability for attack, which imposes a heightened probability that the fundamental interests of T's individual residents t_1-t_n will be contravened.

A useful illustration of this mechanism, and indeed a piece of empirical evidence for its credibility, is the US response to Al-Qaeda following the attacks on September 11. If we plug in the relevant actors from this response to the purported mechanism of risk-imposition, we may simply rewrite the statement above using events that actually transpired. When Al-Qaeda (G), based in a location in Afghanistan (L), coordinated the authorization of military force against residents (s_1-s_n) of the US (S) in 2001, Afghanistan (T), where Al-Qaeda was located, became the target of increased probability for attack, which imposed a heightened probability that the fundamental interests of Afghani residents (t_1-t_n) would be violated. As we now know, the attack *did* occur and many Afghani residents, both military and civilian, *were* killed. Many of these residents had no association whatsoever with Al-Qaeda. Many, in other words, never forfeited their immunity from substantial risk-imposition by attacking, or helping to attack, the US. Yet, these residents suffered at the hands of the US military.

The heightened probability claim begins to explain why groups like the Afghani government might have a responsibility to monopolize decisions of military authorization within their territory. A failure to do so endangers Afghani residents. But while the actions of Al-Qaeda undoubtedly led to this increased risk, two objections immediately arise. The first is that Al-Qaeda did not actually impose risks upon Afghani residents. The US did. Therefore, questions about the responsibility of Al-Qaeda for the individuals who were ultimately harmed—or about the moral failure of the Afghani government to keep Al-Qaeda from imposing risks upon these residents—are the wrong questions to ask. We should probe *US responsibility* if we are concerned with the military and civilian death toll that mounted at the hands of US soldiers. In other words, the heightened probability claim may be true, but so what? While this objection is, I think, powerful, it does not undermine the proposition at hand. The heightened probability claim is simply an empirical premise whose normative import is defended with a second claim (the moral relevance claim). For now, these normative worries may be set aside.

The second objection that the heightened probability claim must confront is an empirical objection. Perhaps the case that I have chosen to illustrate the mechanism of indirect risk-imposition—the US response to Al-Qaeda—is an exception rather than the rule. It may be that states do *not* tend to authorize military force against private actors who authorize military force against them.

While I cannot offer a full rebuttal of this objection, it appears particularly suspect once we recognize that the response-attacks that drive the mechanism of heightened risk may be undertaken by *anyone*. Whether it was the US that authorized military force against Al-Qaeda, the UN Security Council, NATO, or even the Afghani government itself is irrelevant. So long as the private authorization of military force is likely to prompt a military response by *someone*, then the fundamental interests of those who reside in the state from which private military force was launched will be subjected to heightened risk. In order for one to object to the heightened probability claim, one must demonstrate that the probability of military response to private military authorization remains constant after attack—that neither the state that was attacked, the allies of that state, the international community, nor the host-state are any more likely to authorize military force against private authorizing entities than they were before the attack. If the risk of a military response by *any* of these actors increases, the heightened probability claim must be endorsed.

The risk of such a response is enhanced by international laws and norms that increasingly permit states to respond with force abroad to quell private violence, assuming that they respond in states that do not themselves take adequate measures against such violence. The reason is that "failing states who through weakness or ill-will harbor those dangerous to others ... constitute a risk to people everywhere" (ICISS 2001: 5). Sovereignty entails responsibility, as the mantra goes, and the dereliction of responsibility increasingly permits outside military responses.

Note that the ideal piece of data to clinch the heightened probability claim would be a simple proportion of instances in which private actors authorized military force to instances that resulted in a military response. This piece of data could then be compared to the baseline probability of military authorization when private actors have *not* authorized military force. Though no such data exist (to my knowledge), a wealth of data exist on so-called "interstate militarized disputes" in the Correlates of War Project, which are: "united historical cases of conflict in which the threat, display, or use of military force short of war by one member state is explicitly directed towards the government, official representatives, official forces, property, or territory of another state" (Jones et al. 1996: 163). Of the 2,333 militarized interstate disputes that were catalogued from 1816 to 2001, 110 (roughly 5%) ended in war, and hundreds more ended in the use of force short of war (Militarized Interstate Dispute Database 2007). This statistic, of course, severely underestimates the number of instances in which the *authorization of war* by one state would result in the authorization of war by another. Presumably, such a statistic would significantly outpace the 5 percent of militarized interstate disputes that result in war. But even using the more conservative estimate, the probability of states authorizing military force against one another is substantial when the target-state takes hostile military action. Because the baseline probability of military authorization in the absence of military provocation is unknown (and may be too complex for meaningful prediction), we cannot conclude with certainty that a 5 percent chance of war is greater than the usual probability of war. But, given the relative infrequency of war, we should be confident. The question is whether the up-tick in probability will disappear when a private actor is substituted for one of the public actors. And it seems, based on the analysis provided above, that the answer is no.

4.2.2 Alternative international systems

The mechanism that governs heightened probability in the current international system would seemingly be no different (or scarcely different) from the mechanism that governs heightened probability in alternative, even stateless, international systems. If pre-Westphalian history is any indication, groups that are attacked by others are likely to respond with military force (Gat 2006: 147–673). The reason is that such responses are taken to be, and have long been taken to be, the only sure-fire mechanism for self-defense and the deterrence of future attacks.

When military responses are authorized against private actors, even in stateless international systems, these responses will occur in physical space (as in state-based systems), where individuals who are proximate to the attacks become risk-bearers. In addition, groups of individuals with whom the private authorizing entities are associated may be subjected to risk. After all, the formation of such groups, political or otherwise, is certainly possible (and indeed probable) in alternative international systems. The indirect-risk-bearers in a stateless international system—those who are physically proximate to private authorizing entities and who are fellow group members—will likely be affected according to a mechanism that is almost identical to the mechanism that operates in the current international system. Groups that have been attacked are likely to respond with force, whether for reasons of self-defense or punishment (or perhaps other-defense or norm-enforcement). Otherwise, they are likely to expect another group to respond (perhaps the group that "governs" the private actors who attacked). Even in an international system without states, it is difficult to envision a different empirical reality.

4.3   Indirect-risk-bearers: moral relevance

But should such heightened probability entitle affiliates—the residents of the state from which private military force has been authorized (in the current international system) or fellow group members and/or physically proximate individuals (in any international system)—to participate in decisions of private military authorization? An affirmative answer is by no means obvious. First of all, we might think that affected parties possess an entitlement to participate in the decisions of those who have been attacked by private actors, *not* in the decisions of private actors themselves. Perhaps Afghani residents, for instance, merited a say in the US response to Al-Qaeda, not in Al-Qaeda's prior decision to authorize military force against the US. After all, it was the US who ultimately imposed *direct risks* upon Afghanis. Why not simply apply the rationale of the previous section to conclude that Afghanis—more specifically, those Afghanis who meet the criteria regarding immunity and democratic participation—should have been granted input in the US decision to authorize war in Afghanistan? That way, we need not concern ourselves with purported entitlements springing from indirect-risk-imposition. Call this the *alternative-participation objection*.

Second—and notwithstanding the first objection—we might think that, at least in many cases, groups have a responsibility to refrain from responding with military force in the territory from which private actors have launched their attacks. For instance, it may have been that the US had a responsibility to *forego* invading Afghanistan in response to Al-Qaeda's military actions. If so, then Al-Qaeda would not have caused the US to subject Afghanis to *justifiable* risks. Al-Qaeda would have caused the US to subject Afghanis to *unjustifiable* risks. And perhaps private actors ought to have the freedom to authorize military force without consulting indirect-risk-bearers so long as the risks that are ultimately imposed upon these indirect-risk-bearers by third parties would be unjustifiable. Call this the *prohibited-response objection*.

4.3.1 The alternative-participation objection

The problem with the alternative-participation objection is that our fundamental interests are too valuable to demand that the affiliates of private authorizing entities (the Afghanis in this case) forego preventive measures aimed at limiting their risk-exposure by waiting until an invasion of their territory is under consideration. First of all, it may be that participation in the private decision to authorize military force is the *only* way for individuals to protect their fundamental interests. The Afghanis may not have been able to gain a meaningful say in the US decision to invade Afghanistan. Second, even if Afghanis were granted a meaningful say in the US decision to invade Afghanistan, Afghanis may have been able to *more effectively* protect their interests through participation at an earlier stage—in the decision that provoked the US response. If the affiliates of potential private authorizing entities are entitled to participate in a political process with these authorizing entities, all contributors are likely to speak the same language, operate under similar cultural norms, understand relevant political procedures, and so on. This may facilitate a richer protection of one's fundamental interests than participation at a later stage would facilitate (for instance, by Afghanis in US decisions).

Most importantly, the affiliates of private authorizing entities are not likely to know ahead of time whether, or to what extent, they will be able to gain access to later decisions that will affect them. These almost inevitable conditions of uncertainty, coupled with the disastrous consequences of later exclusion, demand sensitivity to the need for anticipatory measures *now*. In particular, we ought not to insist that the affiliates of private authorizing entities sit back and hope that they will gain meaningful participation at a later stage in the risk-imposing process. They ought to be permitted to take reasonable measures of preventive action through participation in decisions of private military authorization at an early stage if the potential risks are sufficiently robust.

Of course, as previously mentioned, when such meaningful participation is not possible, then the affiliates of private authorizing entities *should* get a say in later decisions that directly impose risk upon them. Given the exclusion of so many Afghanis from the political process and the fact that so many did not contribute to the threat posed by Al-Qaeda, many Afghanis *should* have had a say in the US decision to authorize military force against Afghanistan since the risks that were imposed upon them were so pronounced. Fundamental interests are sufficiently valuable to warrant back-up participation, even in multiple decisions, when one is denied opportunities to safeguard these interests.

Jeremy Waldron offers an analogy that is particularly apt here. He (1989: 510) describes a right as imposing not just a single duty but "waves of related duties," which "back it up and root it firmly in the complex, messy reality of political life": given that the interests, which underpin rights, "can be served or disserved" in a variety of ways, "we should not expect to find that *only* one of those ways is singled out and made the subject matter of a duty." As we have already seen in Section 3.4.2, the right to physical security sometimes generates a duty of abstention by the risk-imposer *and* sometimes, according to the all affected fundamental interests principle, a duty to include risk-bearers in decisions of risk-imposition. The point that I offer here, following Waldron, is that *multiple* risk-imposers may become duty-bound to include risk-bearers in decision-making. The all affected fundamental interests principle, in other words, may generate entitlements to participate in waves of successive decisions insofar as these decisions affect our fundamental interests.

4.3.2 The prohibited-response objection

According to the prohibited-response objection, the affiliates of private authorizing entities may be entitled to participate in decisions of private military authorization but *not* when such decisions provoke responses that constitute *unjustifiable* risk-impositions. Consider how the paradigm example explored thus far—the Al-Qaeda attack on the US and the subsequent response by the US in Afghanistan—may have elided a crucial nuance. Many do not fault the US (and certainly the international community did not fault the US) for invading Afghanistan. For most, the private actors (Al-Qaeda) lacked a just cause, and the responder (the US) possessed a just cause. But what if the private actors possessed a just cause, and the responder lacked one? Are we willing to blame the private actor, in possession of a just cause, for not consulting with affiliates? Should we not forego blaming the private actor and simply demand that the responder *refrain from* authorizing military force? If so, then affiliates may only be entitled to participate in private decisions of military authorization (on account of the indirect risks that are imposed) when authorizing entities lack a just cause. Needless to say, this would substantially undermine the force of my argument.

Since the US response to Al-Qaeda is much debated, it will be more fruitful to consider an unequivocal case in which the private actor possesses a just cause and the responder lacks one. That way, we can effectively assess whether the moral relevance of indirect-risk-imposition depends in *any way* upon our background assumptions regarding the justice of the two belligerents—whether, to be more specific, private actors still must include affiliates in decisions of military authorization when they unquestionably possess a just cause and the responder unquestionably lacks one (when the responder certainly *ought not* to respond).

An ideal case for analysis is the one presented in Section 2.5.3, which was given as a critique of Fabre (2012: 142–156; 2014: 401–409) and Lazar (forthcoming: 4–5) for their under-inclusion of risk-bearers. The example, we may recall, consists of a state with three ethnic groups—A, B, and C—each living on separate parcels of land within the state. Group A has long oppressed Groups B and C, and the military of Group A occupies the lands of Groups B and C. Any actor authorizing war against Group A would have just cause to do so, and a private protective agency (PPA_B) in Group B decides to act on this just cause. A response by Group A will be *entirely* unjustified—but will be significant in scale and will target both Groups B and C.[7] As with before, we may assume that the war by PPA$_B$ is proportionate (since the reprisals are unlikely to be catastrophic, though many civilians will be killed). We may also assume that authorization will not be undertaken *on behalf* of Groups B or C, nor would a victory benefit Groups B or C (in the case of a victory, Group D will immediately occupy the territory of Groups B and C, though it will treat the clients of PPA$_B$ more favorably than Group A treated them). Should PPA$_B$ be permitted to launch an attack on Group A without consulting fellow members of Groups B or C?

It seems that PPA$_B$ is *not* justified in authorizing military force against Group A, and the risk-prevention argument has now provided a rationale for precisely why. What matters is not whether private actors authorize military force against an entity that *should* respond but whether private actors authorize military force against an entity that *will* respond. Even if private actors possess a just cause, affiliates may simply not be willing to accept the risks that war entails. The residents of Groups B and C, for instance, may decide, if given the opportunity, that potential retaliation by Group A would make military action unworthy of the risks and therefore ill-advised. For that reason PPA$_B$ may indeed be wronging fellow residents in attacking Group A. Private actors owe it to their affiliates to include them in the risk-imposing decisions of a certain severity even if a response is unjustified (which, of course, is just another way of saying that private actors are entitled to participate in such decisions).

A proponent of the prohibited-response objection may wish to press further, however, and insist that, if affiliates exercise their entitlement of participation and keep private actors from authorizing military force, then the affiliates would now be wronging the private actors. This (perhaps surprisingly) delivers us squarely into the literature on "innocent bystanders."[8] If residents of Group B prevent PPA$_B$ from authorizing military force against Group A (and thus monopolize the decision of military authorization), they would be protecting themselves from risk to their fundamental interests. But the private actor that they block, the objection presses, is innocent in the sense that it has done nothing to make it morally liable to preventive action; the private actor possesses a just cause and is *in the right*. Just as we ask whether individuals are permitted to curtail the liberty of innocent bystanders (and perhaps harm them) to protect themselves from the risk-imposition of an attack, we may ask whether the residents of Group B are permitted to curtail the liberty of (and perhaps harm) innocent private actors to protect themselves from the risk-imposition of an attack. The liberty curtailment (and possible harm) that would be involved is the enforced insistence that such decisions be undertaken collectively.

But the liberty-curtailment imposed on the innocent bystander in this case— the private authorizing entity that possesses a just cause—is mitigated by two factors. First, the force that will be required to prevent private actors from authorizing war will rarely rise to the level of killing. Much of the way that a citizenry will monopolize the authorization of military force will involve intelligence gathering, policing, imprisonment, and other forms of non-lethal force. These non-lethal methods of reserving decisions of military authorization for public bodies are much less problematic than lethal methods in considerations of innocent bystanders.

Second, and perhaps more importantly, the all affected fundamental interests principle allows all individuals, including those wishing to authorize military force, to decide *together* whether the risks of military authorization are tolerable. Shared decision-making appears to be the only fair way to proceed in cases like military authorization. Whether private actors like it or not, their decision to authorize military force may have negative consequences for affiliates. By the same token, the decision by affiliates to take preventive action may have negative consequences for the private actors wishing to authorize military force. A way (perhaps the only way) to fairly resolve these competing positions is to include both sides in a decision about military authorization rather than allow either side to take unilateral action.

Indeed, these considerations point us to a resolution of one difficult challenge that often pops up in discussions of innocent bystanders. While I have portrayed private actors as innocents whose actions may endanger affiliates, I might also have portrayed affiliates as innocents whose actions may endanger private actors. In other words, from the perspective of private actors who are considering the authorization of military force (with a just cause), *the affiliates* are the innocents who may suffer as a result of their efforts to protect themselves. Thus, private actors who wish to authorize military force must take account of affiliates who are innocent bystanders, while affiliates wishing to monopolize military force must take account of private actors who are innocent bystanders. Which side gets to be the one that ultimately thwarts the interests of (and potentially harms) the other (an innocent bystander)? Nozick (1974: 35) first pointed out this tension in his discussion of innocent shields (who I am calling bystanders) and threats:

> If one may attack an aggressor and injure an innocent shield, may the inno-
> cent shield fight back in self-defense (supposing that he cannot move against
> or fight the aggressor)? Do we get two persons battling each other in self-
> defense? Similarly, if you use force against an innocent threat to you, do
> you thereby become an innocent threat to him, so that he may use additional
> force against you (supposing that he can do this, yet cannot prevent his ori-
> ginal threateningness)?

Nozick does not venture an answer to these complicated questions. But one answer, which is applicable to the authorization of military force (though perhaps inapplicable elsewhere), is that innocent bystanders must enter into a democratic process to determine risk-allocation. Since the authorization of military force by a private actor would impose risks upon one innocent bystander (the affiliates who may be attacked in response), and the prevention of this authorization would impose risks upon another (the private actors who believe that military force ought to be authorized), the two must decide together how to proceed.

4.3.3 Moral relevance

I have now challenged two objections to the view that indirect-risk-imposition is excluded from applications of the all affected fundamental interests principle: the alternative-participation objection and the prohibited-response objection. Both objections fail to appropriately acknowledge the ways in which our fundamental interests might go unprotected in the absence of an entitlement to participate in indirect-risk-imposing decisions. The indirect-risk-imposing decision may be the only adequate avenue for our participation. And future direct-risk-imposers cannot always be trusted to include risk-bearers. For these reasons, if we take risks to our fundamental interests seriously, indirect-risk-imposition ought to be sufficient to invoke the all affected fundamental interests principle. In other words, the moral relevance claim must be embraced: the fact that a private decision will not directly impose risks upon the affiliates but will only indirectly impose such risks is not sufficient to eliminate the affiliates' entitlement of participation in that decision.

That the two most compelling objections to the moral relevance claim did not succeed is, upon reflection, perhaps unsurprising. The indirect imposition of risk is really just an imposition of a risk that some further risk will be imposed—it is the risk of a direct-risk-imposition. Consider that, from the perspective of a single Afghani, whenever the US invades Afghanistan, there is a chance of being killed. Let's say that the individual in question faces a 10 percent chance of death as a bearer of *direct-risk* by the US when it invades. But when a private group like Al-Qaeda authorizes military force against the US, there is some probability that the US will respond. Suppose that there was a 50 percent chance that it would invade Afghanistan. The individual in question thus faces a 5 percent chance of death as a bearer of *indirect-risk* by Al-Qaeda when Al-Qaeda attacks the US (a 50 percent chance of invasion by the US, which, if undertaken, imposes a 10 percent chance of death). Because the all affected fundamental interests principle is responsive to *risks*, it would be odd if it were not responsive to both sets of risks—if, in other words, the US was expected to account for the 10 percent chance of death that followed from its decision but Al-Qaeda was not expected to account for the 5 percent chance of death that followed from its decision. Both types of risks, it seems, should be of concern for the all affected fundamental interests principle.

Supposing then that the moral relevance claim is plausible, as I have tried to show, then indirect-risk-bearers, like direct-risk-bearers, may be entitled to participate in private decisions of military authorization. Whether the risks that are imposed upon these risk-bearers (direct and indirect) are of a sufficient severity (and for enough people) to warrant the removal of decisions of military authorization from the private sector is the matter to which we now turn.

4.4 The precautionary principle

If the risks identified in Sections 4.1–4.3 are to ground P₂ of the risk-prevention argument, two questions must be answered. First, is the heightened risk *sufficiently severe* to warrant guaranteed participation in private military authorization (meeting the threshold identified in $P_1 ii$) and to generate a duty of monopolization by those who are affected (as specified in $P_1 iii$)? Second, is the heightened risk spread out across *enough individuals* to require that decisions of military authorization ultimately be withdrawn from the private sector and reserved for public discharge (the demand of $P_1 iv$)? The answer to these two questions, we might say, depends upon the *magnitude* of war. How many people will be harmed, and to what extent, if war occurs? Or, to put these two questions together in a somewhat crude approximation, what will the casualty count be?

One way to demonstrate high risk would focus on both magnitude and probability.[9] A second strategy is available when the harm in question is positively catastrophic, for instance in considerations of climate change and global epidemics, where the result of certain actions (or inactions) may be massive losses of life and widespread suffering even though the probability of such catastrophic losses is low or unknown. According to this second strategy, high risk is demonstrated by focusing primarily on magnitude.

It is the second strategy that the risk-prevention argument employs. The principle upon which my analysis draws is a precautionary principle: when the probability of a catastrophic event is low or unknown, we should err on the side of preventive action rather than inaction. I rely upon the precise formulation of precaution that Henry Shue has offered. According to Shue (2010: 148), when the risk of an event possesses three features, the probability of that event ought to be ignored ("beyond a certain minimum level of likelihood"):

> (1) massive loss: the magnitude of the possible loss is massive, (2) threshold likelihood: the likelihood of the losses is significant, even if no precise probability can be specified, because (a) the mechanism by which the losses would occur is well understood, and (b) the conditions for the functioning of the mechanism are accumulating; and (3) non-excessive costs: the costs of prevention are not excessive (a) in light of the magnitude of possible losses and (b) even considering the other important demands on our resources.

If private military authorization possesses these three features, then we have good reason to insist that such authorization be prohibited and military force publicly monopolized, even when the probability of harm (direct and indirect) is low or unknown.

Before considering whether private military authorization *does* possess these three features, however, I want to offer a brief moral justification of the principle and propose one amendment. The principle, I believe, captures a deeply held, morally defensible, intuition that we tend to universally espouse with respect to loss. As Shue (2010: 148) puts it, there are "losses that would be utterly intolerable, especially 'losses' involving massive deprivations of necessities to which all people, regardless of individual identity, have rights simply as human beings." To frame this point in the language of this chapter, if we are to take our fundamental interests in physical security seriously, we must not allow even a small chance that these interests will be so exhaustively violated in the way that catastrophic events like global warming and unnecessary war entail. This much, I expect, should be uncontroversial. The hard question is whether a given practice, like the private authorization of war, *does in fact* satisfy the conditions that activate the principle.

The one amendment that I wish to suggest before arguing that they do concerns *2b*. Shue elaborated the precautionary principle for application to debates on global warming, where the documented accumulation of greenhouse gasses (the demand of *2b*) helps allay concerns about the unknown probability of catastrophic climate change. But to speak of an accumulation in the context of war—for instance, an arms buildup that must be identified if efforts to prevent private actors from authorizing military force are to be justifiable—is to misunderstand the nature of military monopolization. Individuals can rapidly acquire destructive weapons, and preventive measures are needed to limit private military authorization *ahead* of such buildups. We therefore cannot demand a documented accumulation of arms to defend the practice of public military monopolization (as *2b* demands). Instead, we ought to demand something less stringent.

What I suggest will be dubbed *2b** and will replace *2b*. According to *2b**, actors capable of initiating the conditions for the functioning of the mechanism must be capable of initiating these conditions and have demonstrated a desire to do so. Aside from this replacement, the remaining criteria will be unchanged. Each will be considered in turn. What these considerations show is that the precautionary principle should apply to the direct and indirect-risk-bearers who may suffer as a result of private military authorization.

4.4.1 Massive losses

The *massive losses* of war include death and bodily trauma of every imaginable variety. Aside from such harms to our physical security, wars also cause fear, economic undoing, food shortages, forced migrations, post-traumatic stress, homelessness, and political unrest. Wars, with each passing generation, are increasingly destructive. Four generations after General Sherman characterized wars as hellish, John Hersey (1985: 1, 18) would describe the "noiseless flash" of Hiroshima and the "thick, dreadful miasma" that followed. Two generations after Hersey, a UN High Level Panel (2004: 16) would warn that "with fifty kilograms of highly enriched uranium (HEU), an amount that would fit into six one-liter milk cartons" an individual "could level a medium-sized city." Hundreds of thousands of people may now die horrific deaths at the hands of an adept cadre of private individuals who procure a few milk cartons worth of enriched uranium.[10]

And any military response to such an attack, though hard to envisage, would surely affect countless indirect-risk-bearers. If the US response in Afghanistan to private military authorization by Al-Qaeda is any indication, 2,537 Afghani civilians died in 2009 and 2010 alone, eight years after the response had been authorized—and 5,594 were wounded in these two years (Bohannon 2011). This is not to mention the military personnel who have died in great numbers every year or the deaths and injuries that resulted from war (which is to say that they would never have occurred in the absence of war) but are not part of the official casualty count.

The fact that only some wars unleash the kind of hellishness of September 11 or a nuclear detonation is enough for us to invoke the precautionary principle. The risk-prevention argument aims to show that decisions of military authorization ought to be withdrawn from the private sector and made by public bodies. But, in order to remove particular kinds of decisions of military authorization from the private sector, a public body must *monitor* private decisions of military authorization. In particular, the public body must determine whether a given private decision of military authorization would impose unjustifiable risks for direct and indirect-risk-bearers (and the kind of hellishness of September 11 or nuclear detonation). Only then can it prevent actors from authorizing military force of a particular variety. We might say then that, if there are certain kinds of wars that private actors must not authorize, public bodies must have the final say over which wars qualify. But this just *is* public monopolization over military authorization.

4.4.2 Threshold

The likelihood of massive losses resulting from private military authorization is certainly not known with any great specificity. If estimates were available, the likelihood of these losses, at least of the indirect-variety, would be low. Yet, as *2a* demands, the mechanism of losses that result from private military authorization is very well understood. This mechanism was elaborated in detail in Sections 4.1–4.3. And, as *2b** demands, actors capable of initiating the conditions for the functioning of the mechanism *are* capable of initiating these conditions and have demonstrated a desire to do so. Wars have already been privately authorized that at least have come close to initiating the conditions of massive loss. And these wars have been authorized in an international system of states, which take extensive measures to publicly monopolize the authorization of military force. In a system without such public monopolization, we have reason to believe that privately authorized wars would continue and perhaps proliferate. Moreover, private actors have clearly demonstrated a willingness to use weapons of mass destruction and a capability of doing so; likewise, states around the world have clearly demonstrated a willingness to respond with significant military force. Without public monopolization, the evident desire of private actors to bring about catastrophic destruction and the ever-enhancing technology that enables such destruction should support application of *2b** to private military authorization.

4.4.3 Costs

Finally, the costs of withdrawing decisions of military authorization and reserving them for public discharge are minimal.[11] Unlike the costs of climate-regulation, which must often be borne without immediate benefit, the costs of military monopolization are borne by societies that directly experience the benefits of monopolization. Indeed, the purely short-term financial costs of *not* monopolizing the authorization of military force may be greater than the costs of monopolization, given the severe financial strains of war. This is particularly apparent when we realize that the monitoring required of military monopolization (e.g., intelligence gathering, policing, imprisonment) is a kind of monitoring that domestic law enforcement requires regardless (which societies are likely to institute for reasons that go beyond this book).

4.4.4 Precaution, risk-prevention, and global justice

We may conclude then that the private authorization of military force does meet the criteria for application of the precautionary principle: massive losses, threshold likelihood, and non-excessive costs. Moreover, the precautionary principle itself is quite plausible. Are we to then conclude that the direct and indirect risks imposed by private authorizing entities sit above the threshold that must be met if these decisions are to remain private? I believe, and have tried to show, that we should. Given the horrors of war, the imposition of risk to the fundamental interests of both direct and indirect-risk-bearers simply entails *too much* danger.

Let us take stock then of our analysis. Chapter 3 argued that a decision ought to be withdrawn from the private sector and reserved for public discharge when one course of action under consideration imposes unduly high risks to the fundamental interests of enough individuals (P_1 of the risk-prevention argument). We have now seen in this chapter that the magnitude of potential contraventions to the fundamental interests of direct and indirect-risk-bearers is sufficiently high for enough individuals to place the heightened probability of violation beyond an acceptable threshold of risk (P_2 of the risk-prevention argument). Therefore, decisions of military authorization ought to be withdrawn from the private sector and reserved for public discharge.

At various places in my analysis, I have suggested that entitlements of participation, which are generated by the risk-prevention argument (P_1 and P_2), may demand public international cooperation through institutions like the UN. Given the increasingly menacing capacities of advanced weaponry, individual risk-bearers may be quite geographically isolated from one another. Decisions of military authorization may thus require global input, and it may be that monopolizing institutions must be international in scope. To conclude my presentation of the risk-prevention argument, I wish to reflect briefly on the relationship of my argument to contemporary debates on global justice.

My approach to public monopolization is consistent with many cosmopolitan approaches to international normative theory that seek to capture the enhanced interconnectedness of human interactions. Granted, at least one cosmopolitan theorist, Cécile Fabre (2012: 150–154), explicitly rejects the sort of argument that I give for public military monopolization. Others, including but certainly not limited to Simon Caney (2005: 205–207), David Rodin (2002: 174–188), Fernando Tesón (2003: 93–129), and James Pattison (2014: 205–220), embrace some of the conclusions of the risk-prevention argument via alternative arguments. But it seems that cosmopolitan thinkers should be quite receptive to the potentially global reach of the risk-prevention argument, especially since many have explicitly endorsed the all affected interests principle (Pogge 2002: 190; Held 2004: 100; Cohen and Sabel 2006: 169).[12]

Christian List and Mathias Koenig-Archibugi (2010: 76) eloquently summarize the relationship between global interconnectedness and international cooperation: "The world is increasingly characterized by transnational interdependence, cross-border policy externalities, and the widely perceived need to provide global collective goods and to avoid global collective bads." They cite, as examples, greenhouse gas emissions, the flow of refugees across borders, and the spread of rapidly lethal infectious disease—which all demand global governance. My analysis in Part I may simply add the authorization of war to this list of global collective bads. The effects of war, like climate change, refugee migrations, and infectious diseases, tend to ripple across the globe now more than ever (and certainly in the previous century more than all prior centuries). As a result, they demand, at a minimum, more *public international cooperation* and thus more robust efforts aimed at public international monopolization.

4.5 Public military authorization and compensation

The conclusion of the risk-prevention argument, as we have seen, is that public bodies ought to monopolize the authorization of military force to ensure that private actors do not impose undue risks upon one another. We have not yet seen, however, whether public bodies, when making decisions of military authorization (decisions that they alone are permitted to make) are ever justified in deciding positively to authorize military force. Pacifists who embrace public monopolization would insist that they do not. Such pacifist positions distinguish between the policing required of public monopolization and the military deployment required of public authorization (Teichman 1986: 38–45). This distinction has been pressed in my own analysis and so will not be elaborated further.[13]

From this distinction, a pacifist extension of the risk-prevention argument may be constructed. The pacifist, like the just war theorist, may demand the public policing of military monopolization in light of the risks to which direct and indirect-risk-bearers are subjected by private decisions of military authorization. But the pacifist may insist further that *no* decisions of military authorization, whether public or private, are in fact capable of adequately reflecting the preferences of risk-bearers. Strengthened international institutions, the pacifist might claim, could successfully enable the satisfactory representation of risk-bearers in war. But current institutions do not. Thus, wars, whether public or private, may not be justifiably authorized *now*. If, in the future, it turns out that strengthened international institutions are capable of better representing risk-bearers, and wars are perhaps rendered more like acts of domestic policing (which pacifist advocates of public military monopolization embrace), *then* public military authorization may be justifiable.

But a careful reflection on the implications of the risk-prevention argument indicates that, *contra* the pacifist endorsement of A_0, wars must be justifiable now, even in the absence of strengthened global institutions. When a public monopolizing entity is faced with a threat, the entity must make a decision about *competing risks*. Were the government to refrain from authorizing military force, those who have been disarmed may be exposed to acute risks on a massive scale—say, at the hands of an invading army. Yet, were the government to authorize war, it would expose others to substantial direct-risks and may perhaps expose those who have been disarmed to substantial indirect-risks.

Supposing that risk-bearers have been properly consulted, *sometimes* a government faced with this choice must be entitled to authorize military force for the very reason that military force must be monopolized in the first place: to prevent un-chosen risk. The proper weighting of competing risks will never be easy. But there will be times when the risks that are imposed by the authorization of military force are *much less* than the risks that are imposed by holding fire. Public actors must not be denied the opportunity to authorize military force in these instances, for such a denial would subject those who have been disarmed to risks in precisely the way that I have sought to combat.[14]

Another way to frame the position that I take here is in terms of compensation. Individuals who have been disarmed by public monopolizing entities must be compensated in some way. As Nozick (1974: 114) puts it: "those who act in self-protection in order to increase their own security" must "compensate those they prohibit from doing risky acts which might actually have turned out to be harmless for the disadvantages imposed upon them." One form of compensation, which is preferred by Nozick, is protection in kind. Those who disarm individuals to prevent them from engaging in risky behavior ought to provide them with protective services. Nozick's argument is pitched at private protective agencies in a state of nature, but it may also be pitched at public actors. When public actors prevent private actors from protecting themselves on the grounds that these private actors expose others to risk through such efforts at protection, then the public actors incur a duty to prevent others (e.g., enemies in war) from exposing them to risk.

The compensation argument, as we can see, ultimately gains its force from the same principles that underlie the risk-prevention argument. According to the risk-prevention argument, the reason why decisions of military authorization ought to be withdrawn from the private sector is that the fundamental interests of risk-bearers can only be protected in this way. But if, after withdrawing decisions of military authorization from the private sector, a public body were barred from authorizing military force itself, the fundamental interests of those who have been disarmed would be left vulnerable. The very considerations that motivated us to advocate military monopolization must motivate us to support a permission of military authorization. Whether a particular war ought to be authorized will depend upon the risks of that war and the risk-tolerance of those who will likely be affected. But public bodies owe it to those who have been dis-armed to authorize military force when the fundamental interests of these indi-viduals are threatened and the risks of war are tolerable to them. To deny this would be to deny the principle upon which the risk-prevention argument was based: individuals must be able to protect their fundamental interests.

4.6 Emergency authorization

We have now seen that public actors have a duty to monopolize the authoriza-tion of military force (according to the risk-prevention argument) and a duty to authorize military force when those who have been disarmed are threatened (according to the compensation argument). But these duties are limited. The first limit to note is that individuals cannot be expected to monopolize military force when they are *incapable* of the institutional steps needed to prevent private actors from authorizing military force. The attribution of a duty requires, at least, that the subject of the duty be capable of discharging that duty. To return again to the Warsaw Ghetto example first described in Section 2.4.2, if the 400,000 Jews residing in this ghetto could not communicate, much less coordinate, with one another, then the putative duty to collectively monopolize force would be moot.

A second limit to note is that, when military force is not publicly monopo-lized, the duty of public military authorization, as derived from the compensa-tion argument, dissolves. The compensation argument calls for protection of those who have been disarmed, but we can demand no such protection according to the compensation argument when no individuals have been disarmed. Thus, even if Warsaw Ghetto residents were capable of military authorization but were *in*capable of military monopolization, they would lack a duty of public military authorization according to the compensation argument.[15] As we know, Warsaw Ghetto residents *were* capable of coordinating with one another and did in fact stage an uprising (Guttman 1994). But the point remains that duties of collective monopolization and authorization abate insofar as opportunities for coordination are quashed.

The difficult question is, if communication within the ghetto had been thor-oughly suppressed, would individuals (or groups of individuals) have been per-mitted to authorize military force independently, without consulting risk-bearers? I suggested in Section 2.4.2 that Warsaw Ghetto residents may *not* have pos-sessed a right to privately authorize military force, because institutional meas-ures were available for residents to transform private decisions of military authorization into public decisions. The risk-prevention argument has now pro-vided a framework to explain why democratic deliberation must be pursued in the ghetto—this is how individuals can (and must, when possible) protect their fundamental interests in physical security.

But the risk-prevention argument also provides a rationale for why democratic deliberation is not the only available method for collective decision-making. Reliance upon secondary indicators of presumptive consent (polls, rallies, media, past discussions) may also suffice when democratic deliberation is impossible. The reason is that such indicators may serve as *valid proxies* for genuine preferences. The risk-prevention argument is ultimately concerned that decisions of military authorization reflect the preferences of risk-bearers. To see how secondary indicators of presumptive consent may reliably demonstrate true preferences, we need only consider a simple hypothetical variation on the Warsaw Ghetto example. In this variation, a single resident positioned near an underground tunnel receives a large shipment of weaponry. She is unable (like all ghetto residents) to consult with others about war. But she is immensely confident that *all* immune risk-bearers, including Poles beyond the ghetto walls, desire military authorization, because, prior to Germany's invasion of Poland, a newspaper-survey of *all* Warsaw residents found that, in a hypothetical scenario like this one, 100 percent favored military authorization. In this case, the would-be authorizing agent is acting on her best interpretation of collective preferences, and the interpretation itself is based on an abundance of sound evidence. We ought to insist that she directly consult with risk-bearers at the earliest possible moment (ceding all decision-making authority at that point to the collective) and that, following Fabre (2012: 155), she help establish institutional structures to ensure that she is held accountable and possibly punished after authorizing war. But military authorization undertaken according to these requirements in situations of emergency is *public* in the relevant sense (of both attempting to reflect and submitting to collective preferences).[16] And it is consistent with requirements of the risk-prevention argument.

Of course, this is an exceedingly artificial example. Would-be authorizing entities often lack reliable indicators of presumptive consent. When an agent is not acting on the preferences of the collective—because she is not sufficiently *sure* about these preferences—then she is engaged in private military authorization and seemingly violates the risk-prevention argument. What do we say in these circumstances? I think that we must say that military authorization is *unjustifiable*. The risk-prevention argument is premised on a deep intuition that we ought not to impose un-chosen risks upon the fundamental interests of those who have done nothing to forfeit their freedom from substantial risk-imposition. When the preferences of risk-bearers cannot be known with reliable accuracy, wars ought not to be authorized. This means that individuals must sometimes die in order to avoid imposing unjustifiable risks upon others.

While this conclusion entails a form of pacifism that may deviate from considered judgments about war, it should resonate with considered judgments about individual self-defense. When the victim of an attack can save herself only by shooting an innocent bystander—say, the wife of her attacker in order to distract him—the victim may be prohibited from undertaking self-defense.[17] "Even under the shadow of death we recognize a duty to discriminate and not to harm certain classes of person" (Rodin 2002: 81), and one such class is innocent

bystanders.[18] Likewise, we insist that the victim refrain from subjecting the innocent bystander to certain forms of risk. It does not become permissible for the
victim to fire at an innocent bystander if she is a less-than-perfect shot (with a
10% likelihood of missing) or if the gun is a six-chamber revolver with only five
bullets loaded (and the victim has time for one shot). Nor is it always permissible for the victim to shoot at the *assailant* knowing that the bullet has a 90
percent chance of killing the innocent bystander or a 90 percent chance that it
will startle the assailant such that he detonates a blast killing 100 innocent
bystanders. Even those who embrace the doctrine of double effect—and thus
insist that some harms, though impermissible to inflict intentionally, may be
inflicted as foreseen side effects of action—recognize that foreseen side effects
may be disproportionate to the harm inflicted.[19] If so, an individual must refrain
from inflicting the harm that will collaterally kill bystanders *even if* they must
die as a result. The simple point is that there are *some* instances in which individuals must forego self-defense rather than bring about the deaths of others
(even when doing so is merely foreseen, not intended). Given the potentially
horrific losses of war, *military* authorization is surely one such example.

One may be tempted, at this stage, to raise two related objections. The first is
that the risk-prevention argument wrongfully *assumes* wars to be inevitably
risky. Perhaps if the risks of war were sufficiently low, then individuals may be
justified in private military authorization. We could imagine a scenario in which
the weaponry shipped through the underground tunnel in the Warsaw Ghetto is
both exceptionally powerful and precise (or a scenario in which a wealthy for-
eigner from outside the ghetto funds an exceptionally powerful and precise
humanitarian intervention). The overwhelming force would incapacitate all
German aggressors (hence neutralizing indirect risk), and its precision would
spare all who are immune, both within the ghetto and beyond (hence side-
stepping direct risk).

The problem, first of all, is that, if such minimally risky wars were possible,
the bar to demonstrate presumptive consent via secondary indicators would be
exceedingly low. Most individuals (to understate the point) would welcome
efforts to save them if these efforts imposed minimal risk. I do not mean to
suggest, of course, that no bar exists. If the Warsaw Ghetto residents expressed a
clear preference *not* to be saved via a largely risk-free war, then the strike would
be unjustifiable. But, where authorizing entities cannot consult with risk-bearers,
the threshold of evidence that they must marshal to demonstrate presumptive
consent will be lower than the threshold for risky wars.

Far more importantly, and the first point notwithstanding, wars *are* risky. We
have yet to observe a war in human history that did not impose substantial risks
on the fundamental interests of *many* non-liable individuals. This is how wars
differ from individual cases of self-defense. Unlike the scenarios above in which
an individual opens fire on her assailant knowing that she will subject one innocent bystander to risk (direct risk) or that she will startle her assailant to detonate
a blast killing 100 innocent bystanders (indirect risk), wars kill thousands, or
tens of thousands, or millions of innocent people. *If* we come to live in a world

where wars are largely risk-free (though I doubt that such a world exists), then the risk-prevention argument loses its force. It is contingent upon the claims defended in Sections 4.1–4.3 that wars impose considerable risks on *numerous* non-liable individuals. The risk-prevention argument is impotent against risk-free wars, because there are no risk-bearers to motivate the argument. But so long as we continue to live in the world that we do, wars will remain risky endeavors, and the risk-prevention argument will apply.[20]

A second related objection is that, while wars may indeed be risky, the risk-prevention argument is simply, at its core, an argument about proportionality, not legitimate authority. Would individuals not be justified, the objection might press, in privately authorizing war so long as it was (simply) proportionate? In such cases, individuals *could* impose risks upon others *if* they were proportionate, and the risk-prevention argument would seemingly add little to the ethics of war.

But, as I have highlighted, a private war may be *proportionate* but still intolerable to the risk-bearers. Given the magnitude of wrongs committed by the Nazi regime, nearly any war initiated by Warsaw Ghetto residents (or by others for their sake) would likely be proportionate. And, yet, this tells us nothing about whether members of the Warsaw Ghetto would *support* a war given the risks. The residents may object on a variety of grounds, and these residents ought not to be subjected to the (albeit proportionate) risks of war that they do not judge to be tolerable.

4.7 Objections to the risk-prevention and compensation arguments

Having now seen then that private military authorization is unjustifiable, even in emergency circumstances, I will briefly consider and rebut three final objections to the position that I have established.

4.7.1 Which collective?

The first, given by Fabre (2012: 151–152), is that the risk-prevention argument "proves too much" by insisting that "any major decision affecting a community's future should be made with the consent (direct or indirect) of the people"; it implies that "the decision to constitute oneself into a self-determining community is *itself* unjust, since prior to that decision being made, the community is not (yet) self-determining." But, of course, the risk-prevention argument does not insist that "any major decision" require collective decision-making, only decisions that affect fundamental interests. And the decision to constitute oneself into a collective may not affect the fundamental interests of others. Moreover, the all affected interests principle, which underpins the risk-prevention argument, provides a ready answer to the problem of constituting the collective; the collective must be constituted by all who are affected by its decisions. Only *then* is democratic decision-making possible. Fabre's critique has simply restated the boundary problem, and the all affected interests principle offers a solution to that problem.[21]

4.7.2 Duty of authorization

A second objection offered by Fabre (2012: 152–153) is that the risk-prevention argument must not apply when authorizing entities have a clear just cause for war. She gives the counterfactual example of President Roosevelt authorizing war after the bombing of Pearl Harbor *if* he had "dispensed with the approval of Congress" and "indeed, with the approval of the American people"; Fabre (2012: 153) argues that "[i]f any decision to go to war is to be deemed just, this one surely is." But, the risk-prevention argument has demonstrated that even just wars may be intolerable to risk-bearers and therefore unworthy of being fought. Entry into a world war seems to be no exception.

But suppose that the objection is not that any wars with a just cause may be privately authorized but that wars with a just cause that individuals have a duty to prosecute (e.g., humanitarian interventions) may be authorized privately. Lazar (forthcoming: 15–16) levels this objection, arguing that public authorization may not be required when the collective decides against war "but its members actually have a duty to contribute to fighting the war." Reitberger (2013: 76) describes the Swedish Volunteer Corps of individuals who came to Finland's defense in 1939–40 in precisely this way: "When the Swedish state did not actively intervene on Finland's behalf" Swedish volunteers "organized a private army to fight in the war, albeit under Finnish command and with plenty of support (although not 'authorization') from the Swedish government." Is the private authorization of war not justifiable (and even obligatory) in these cases? The key point to draw out is that individuals may fulfill their humanitarian obligations by fighting in a *publicly authorized* war, as Swedish citizens did. If there is no publicly authorized war, then private actors may seek direct consultation with risk-bearers to undertake collective authorization *together* or rely upon secondary indicators of presumptive consent when such consultation is not possible to authorize war *on their behalf*. But, when private actors possess humanitarian duties, yet lack sufficient confidence that the beneficiaries of these duties (and other risk-bearers) desire intervention, then private actors ought not to act.

4.7.3 Private philanthropy

A final objection to my position is that it cannot prohibit putatively private actors from authorizing military force when they *do* have the endorsement of risk-bearers.[22] A philanthropist who wished to stop genocide, for instance, might consult with relevant risk-bearers, whether directly or through secondary measures of presumptive consent, and hire a PMSC like Blackwater to stage an intervention if approved by the collective. Such military authorization, it would seem, would be private and yet also justifiable according to the risk-prevention argument.

However, as we saw in Section 3.4.3 (the defense of P_1iii), risk-bearers have an obligation to *ensure* that they exercise a final say over decisions of military authorization, and private actors have an obligation to submit to collective enforceable oversight. The decision of military authorization, in other words, must be reserved for the collective and, in that sense, must remain public. If the collective chooses to accept philanthropic donations to fund its war, the collective would nevertheless be prosecuting a publicly authorized war, which need not fall foul of the risk-prevention argument. If, by contrast, the philanthropist were to authorize war apart from the collective, then military action would violate the risk-prevention argument and would therefore be unjustifiable.

4.8 Conclusion

In this chapter I have demonstrated that public entities must monopolize the authorization of military force (*risk-prevention argument*) and that these entities must authorize military force when the individuals who have been disarmed are threatened (*compensation argument*). In addition, I have objected to several lines of criticism that could potentially undermine my arguments. This chapter therefore stands as a rejection of A_0, A_1, and A_2.

Notes

1　See Section 3.2 for the distinction between direct- and indirect-risk-bearers.
2　From here forward, when I refer to "traditionalists," I refer to those who accept the moral equality of soldiers and non-combatant immunity (as conventionally rendered). When I refer to "revisionists," I refer to those who reject the moral equality of soldiers and non-combatant immunity (as conventionally rendered).
3　See McPherson (2007: 529), Finlay (2013: 151), Lazar (2010: 210; forthcoming: 9), and Fabre (2012: 82) for a similar point. McPherson (2007: 529) cites this sobering statistic about what he calls the "brute reality of war for non-combatants": between 1900 and 1990, there were 43 million combatant casualties in war and 62 million civilian casualties.
4　Needless to say, we are not interested in *legal* immunity but in *moral* immunity. The fact that Al-Qaeda is a non-state actor that does not qualify as a legal belligerent should not distract us.
5　This was precisely the upshot of the arguments considered in Section 3.2 that were offered by McMahan (2010: 47–54) and Pattison (2010: 142–144) in the context of humanitarian intervention. During humanitarian crises, individuals in the target-state are shut out of the political process by their own government as the government commits human rights violations against them. These individuals, who become the direct-risk-bearers of humanitarian intervention, *should be* entitled to participate in decisions of humanitarian military authorization according to McMahan and Pattison. Here, I simply push the point further to suggest that direct-risk-bearers may be entitled to participate even in non-humanitarian military authorizations that may expose them to risk.
6　In Section 4.2, we will see that the indirect-risk-bearers who may be entitled to participate in decisions of military authorization are numerous just like direct-risk-bearers. This only amplifies the point now under consideration.
7　Some just war traditionalists may be unwilling to grant that a response by Group A *would be* unjustified. Perhaps Group A is *justified* in responding. To proponents of

such a position, however, my case for indirect-risk-imposition will be even stronger. Private actors who authorize military force with a just cause cannot bank on the fact that the enemy will refrain from responding. It is therefore particularly imperative that such private actors consult with affiliates.

8 On "innocent bystanders" and "innocent threats," see especially Nozick (1974: 34–36), Thomson (1991), Uniacke (1994: 156–231), and Rodin (2002: 79–83).

9 Risk should be understood as a magnitude multiplied by a likelihood. More specifically, it is the "statistical *expectation value* of an unwanted event which may or may not occur" where the expectation value is "the product of its probability and some measure of severity" (Hannsson 2014: 2).

10 For a powerful depiction of this threat, see Allison (2004).

11 I am assuming the existence of well-functioning public institutions here. In Section 4.6, I will examine emergency circumstances in which well-functioning public institutions do not exist.

12 Note, however, that application of the risk-prevention argument does not depend upon the application of comprehensive principles of distributive justice to the global realm. It depends only upon more minimalist applications of justice. Remember that we are interested in fundamental interests in physical security, which are protected by rights. Consequently, the potentially global push of the risk-prevention argument should not trouble those who insist upon the presence of states for the application of comprehensive principles of distributive justice. For such a "statist" view, see Nagel (2005).

13 See, in particular, Section 1.7 and Section 2.2.

14 If those who have been disarmed collectively *choose* to forego military authorization, then, of course, they may forego military authorization.

15 Of course, these individuals may have possessed a duty of military authorization on some other grounds besides the compensation argument.

16 See Section 2.4.2 for the definitional distinction between "public" and "private" military authorization.

17 See Rodin (2002: 81) and Nagel (1979: 69) for this example.

18 More difficult cases involve "innocent threats" and "innocent aggressors." The classic example of the former is Nozick's (1974: 34) fat man—who has been thrown down a well and will crush (and kill) an individual at the bottom (though the fat man himself will survive) unless the individual at the bottom zaps the fat man with a ray-gun. The classic example of the latter is someone who has been injected with a drug impairing his judgment and causing him to kill through no fault of his own (Thomson 1991: 284). A prohibition on killing "innocent threats" and "innocent aggressors" in self-defense is far more controversial than killing "innocent bystanders."

19 On the doctrine of double effect, see especially Foot (2002 [1967]: 19–31), Quinn (1989), Kamm (1991), McMahan (1994), McIntyre (2001), Walzer (2006a: 151–159), and Masek (2010).

20 We should acknowledge, of course, that an individual in the Warsaw Ghetto who acquires weapons may be permitted to use force short-of-war to defend herself (though not if this would pose risks to others above a certain threshold).

21 See Section 3.3 for an explanation of the boundary problem.

22 I thank James Pattison for pressing me on this point.

References

Allison, Graham (2004). *Nuclear Terrorism: The Ultimate Preventable Catastrophe* (New York: Times Books).

Bohannon, John (2011). "Counting the Dead in Afghanistan," *Science*, Vol. 331, No. 6022, pp. 1256–1260.

Caney, Simon (2005). *Justice Beyond Borders: A Global Political Theory* (Oxford: Oxford University Press).

Cohen, Joshua and Charles Sabel (2006). "Extra Rempublicam Nulla Justitia?," *Philosophy and Public Affairs*, Vol. 34, No. 2, pp. 147–175.

Fabre, Cécile (2012). *Cosmopolitan War* (Oxford: Oxford University Press).

Fabre, Cécile (2014). "Rights, Justice, and War: A Reply," *Legal Theory*, Vol. 33, No. 3, pp. 391–425.

Foot, Phillipa (2002 [1967]). "The Problem of Abortion and the Doctrine of Double Effect," *Virtues and Vices: An Other Essays in Moral Philosophy*, pp. 19–31.

Gat, Azar (2006). *War in Human Civilization* (Oxford: Oxford University Press).

Gutman, Israel (1994). *Resistance: The Warsaw Ghetto Uprising* (New York: Houghton Mifflin).

Hansson, Sven Ove (2014). "Risk," in Edward N. Zalta (ed), *The Stanford Encyclopedia of Philosophy* (Spring 2014 Edition). Available at http://plato.stanford.edu/entries/risk/ (accessed December 28, 2015).

Held, David (2004). *Global Covenant: The Social Democratic Alternative to the Washington Consensus* (Cambridge: Polity Press).

Hersey, John (1985). *Hiroshima*, second edition (New York: Alfred A. Knopf).

International Commission on Intervention and State Sovereignty (ICISS) (2001). *The Responsibility to Protect* (Ottawa: International Development Research Center). Available at http://responsibilitytoprotect.org/ICISS%20Report.pdf (accessed December 28, 2015).

Jones, Daniel, Stuart A. Bremer, and J. David Singer (1996). "Militarized Interstate Disputes, 1816–1992, Coding Rules and Empirical Patterns," *Conflict Management and Peace Science*, Vol. 15, No. 2, pp. 163–213.

Kamm, Frances (1991). "The Doctrine of Double Effect: Reflections on Theoretical and Practical Issues," *Journal of Medicine and Philosophy*, Vol. 16, No. 5, pp. 571–585.

Lazar, Seth (2010). "The Responsibility Dilemma for *Killing in War*: A Review Essay," *Philosophy and Public Affairs*, Vol. 38, No. 2, pp. 180–213.

Lazar, Seth (forthcoming). "Authorisation and the Morality of War," *Australasian Journal of Philosophy*, pp. 1–19.

List, Christian and Mathias Koenig-Archibugi (2010). "Can There Be a Global Demos? An Agency-Based Approach," *Philosophy and Public Affairs*, Vol. 38, No. 1, pp. 76–110.

Masek, Lawrence (2010). "Intentions, Motives, and the Doctrine of Double Effect," *The Philosophical Quarterly*, Vol. 60, No. 240, pp. 567–585.

McIntyre, Alison (2001). "Doing Away with Double Effect," *Ethics*, Vol. 111, No. 2, pp. 219–255.

McMahan, Jeff (1994). "Revising the Doctrine of Double Effect," *Journal of Applied Philosophy*, Vol. 11, No. 2, pp. 201–212.

McMahan, Jeff (2004). "The Ethics of Killing in War," *Ethics* Vol. 114, No. 4, pp. 693–733.

McMahan, Jeff (2006a). "On the Moral Equality of Combatants," *The Journal of Political Philosophy*, Vol. 14, No. 4, pp. 377–393.

McMahan, Jeff (2006b). "The Ethics of Killing in War," *Philosophia*, Vol. 34, No. 1, pp. 23–41.

McMahan, Jeff (2008). "The Morality of War and the Law of War," in David Rodin and Henry Shue (eds), *Just and Unjust Warriors: The Legal and Moral Status of Soldiers* (Oxford: Oxford University Press), pp. 19–43.

McMahan, Jeff (2009). *Killing in War* (Oxford: Oxford University Press).

McMahan, Jeff (2010). "Humanitarian Intervention, Consent, and Proportionality" in N. Ann Davis, Richard Keshen, and Jeff McMahan (eds), *Ethics and Humanity: Themes from the Philosophy of Jonathan Glover* (Oxford: Oxford University Press), pp. 44–72.

McPherson, Lionel (2007). "Is Terrorism Distinctively Wrong?," *Ethics*, Vol. 117, No. 3, pp. 524–546.

Militarized Interstate Dispute Database (2007). "The MID3 Data Set, 1816–2001." Available at www.correlatesofwar.org/data-sets/MIDLOC/mid-location-1816-2001-v1-1 (accessed December 28, 2015).

Nagel, Thomas (1979). "War and Massacre," in *Mortal Questions* (Cambridge: Cambridge University Press).

Nagel, Thomas (2005). "The Problem of Global Justice," *Philosophy and Public Affairs*, Vol. 33, No. 2, pp. 113–147.

Nozick, Robert (1974). *Anarchy, State, and Utopia* (New York: Basic Books).

Pattison, James (2010). *Humanitarian Intervention and the Responsibility to Protect* (Oxford: Oxford University Press).

Pattison, James (2014). *The Morality of Private War* (Oxford: Oxford University Press).

Pogge, Thomas (2002). *World Poverty and Human Rights: Cosmopolitan Responsibilities and Reforms* (Cambridge: Polity Press).

Quinn, Warren (1989). "Actions, Intentions, and Consequences: The Doctrine of Double Effect," *Philosophy and Public Affairs*, Vol. 18, No. 4, pp. 334–351.

Rodin, David (2002). *War and Self-Defense* (Oxford: Oxford University Press).

Shue, Henry (2010). "Deadly Delays, Saving Opportunities," in Stephen Gardiner, Simon Caney, Dale Jamieson, and Henry Shue (eds), *Climate Ethics: Essential Readings* (Oxford: Oxford University Press), pp. 101–111.

Teichman, Jenny (1986). *Pacifism and the Just War* (Oxford: Basil Blackwell).

Tesón, Fernando (2003). "The Liberal Case for Humanitarian Intervention," in J.L. Holzgrefe and Robert O. Keohane (eds), *Humanitarian Intervention: Ethical, Legal, and Political Dilemmas* (Cambridge: Cambridge University Press), pp. 93–129.

Thomson, Judith Jarvis (1991). "Self-Defense," *Philosophy and Public Affairs*, Vol. 20, No. 4, pp. 283–311.

Uniacke, Susan (1994). *Permissible Killing* (Cambridge: Cambridge University Press).

United Nations (1945). *Charter of the United Nations*. Available at www.un.org/en/documents/charter/index.shtml (accessed December 28, 2015).

United Nations High Level Panel on Threat, Challenges and Change (2004). *A More Secure World: Our Shared Responsibility* (New York: United Nations Department of Public Information). Available at www.un.org/en/peacebuilding/pdf/historical/hlp_more_secure_world.pdf (accessed December 28, 2015).

Waldron, Jeremy (1989). "Rights in Conflict," *Ethics*, Vol. 99, No. 3, pp. 503–519.

Walzer, Michael (2006a). "Regime Change and Just War," *Dissent*, Vol. 53, No. 3, pp. 103–108.

Walzer, Michael (2006b). *Just and Unjust Wars: A Moral Argument with Historical Illustrations*, 4th edition (New York: Basic Books).

Walzer, Michael (2006c). "Response to McMahan's Paper," *Philosophia*, Vol. 34, No. 1, pp. 43–45.

Part II
Supplying war

5 Governance

Part I asked whether private military *authorization* is morally justifiable and argued that it is not. Public actors must prevent private actors from authorizing military force (monopolization) and use military force when the circumstances demand (authorization). Part II now turns to the question of whether private military *supply* is morally justifiable. Recall the difference between authorization and supply. An entity that undertakes military authorization decides both that military force will be supplied and by whom, and an entity that undertakes military supply then executes the various tasks that have been authorized for performance.[1] Whereas Part I delivered a conclusion about the role of private actors in initiating wars, Part II will deliver a conclusion about the role of private actors in fighting wars.

This book began with six possible arrangements for the societal provision of military force: A_0–A_5. Because A_1 and A_2 both include private military authorization and because Part I offered a rejection of private military authorization, A_1 and A_2 were eliminated as potentially justifiable arrangements. Moreover, because A_0 entails *no* military authorization and Part I defended the authority of public actors to authorize military force, A_0 was eliminated as a potentially justifiable arrangement. As we move into Part II, the list of defensible candidates for military provision thus stands at three arrangements: A_3, A_4, and A_5. The aim of Part II is to appraise these arrangements and, in so doing, round out a theory of military privatization.

The characteristic of each arrangement A_3–A_5 that distinguishes one from the next is the place afforded to private military supply. While public actors authorize wars under all three arrangements, the militaries upon which these public actors rely in A_3–A_5 differ: under A_3, publicly controlled militaries are composed entirely of private personnel;[2] under A_4, publicly controlled militaries are composed of some private personnel and some public personnel;[3] and under A_5, publicly controlled militaries are composed entirely of public personnel. To assess the moral merits of each arrangement, we must develop an account of which military functions ought to be publicly undertaken and which may be justifiably outsourced. Such an account enables judgments about the responsibilities of a society in instituting one arrangement over another.

Part II argues that publicly controlled militaries may *not* justifiably outsource all military functions to the private sector. In particular, according to the governance argument (Chapter 5) and punishment argument (Chapter 6), a government may not privatize its *military leadership*, taken to be that group of military personnel who serve as commanders or their superiors on the chain of command. According to the control argument (Chapter 7), a government may privatize its *rank-and-file personnel*, taken to be that group of military personnel outside of a military's leadership, but only if they do so under a unified chain of command. The chapters of Part II place a number of restrictions on military outsourcing, which, if enshrined, would substantially limit the practice.

This chapter defends the governance argument. According to the governance argument, some officers exercise substantial power over civilian decisions of military authorization and supply, and they make weighty decisions in battle that affect the fundamental interests of those whom they protect and those whom they attack. For these reasons, I argue that the responsibilities of high-ranking military officers must be publicly discharged. The chapter proceeds as follows. First, I define the suppliers that animate Part II: *independent military contractors* and *private military and security companies* (PMSCs). Next, I consider and challenge a prominent defense of private armies that is given by Cécile Fabre (2010; 2012: 208–238).[4] According to her defense, all military functions may be outsourced to the private sector. After criticizing Fabre, I propose the governance argument in the third section as a means of limiting the scope of permissible military outsourcing. The fourth and fifth sections then offer two independent rationales for the conclusion of this argument. If my analysis is sound, we may conclude that private armies are objectionable and that the responsibilities of high-ranking military personnel ought not to be privatized.

5.1 Independent military contractors and PMSCs

All private personnel who supply military services to entities that have authorized military force are referred to as *military contractors*.[5] More specifically, to borrow from Fabre (2012: 210), military contractors are taken to be those individuals who supply military services to "belligerent[s], against payment, outside the state's military recruitment and training procedures, either directly to a party in a conflict, or through an employment contract" with a private firm.[6] Contractors perform a range of functions: interrogating prisoners, building bases, delivering ammunition, killing enemy combatants, flying sorties and, in general, providing whatever services are required by militaries for success in war. On my account, however, they do not include those who simply supply *materials* (food, weapons, uniforms, and so on) to militaries.[7] Instead, military contractors supply military *services* to militaries.[8]

Military contractors, for the purposes of this book, fall into one of two classes: *independent contractors* and *non-independent contractors* (the latter of which will be synonymous with *PMSC employees*). The difference between independent contractors and non-independent contractors is that the former contract directly with the entity that has authorized war, while the latter serve as paid members of organizations, which, in turn, contract with the entity that has authorized war.[9] A less formal way of understanding this distinction is that independent contractors receive their paychecks from the belligerents whom they serve while non-independent contractors receive their paychecks from organizations that contract with belligerents.

We should be careful to note that a PMSC that supplies military services for financial payment to belligerents is just "a collection of individuals" that supplies military services for financial payment to belligerents. PMSCs, on my account, include corporations (for-profit or not-for-profit), partnerships, and any other enterprise of more than one individual that supplies military services to belligerents for payment.[10] This means that terrorist organizations, rebel groups, and other paramilitary outfits paid by a belligerent to provide military services in war may carry the label of a PMSC in my analysis. The *only* difference between a PMSC and an independent military contractor is that the former is a group of individuals while the latter is a single individual. Because non-independent contractors include all (and only those) who are employed by PMSCs, all non-independent contractors are PMSC employees. As such, all military contractors are either classified as independent contractors or PMSC employees.

Other political theorists, including both critics of privatization like Cheyney Ryan (2009) and those more amenable to the practice like Fabre (2010; 2012: 208–238), opt to label all private participants on the battlefield as mercenaries. But considerations of ordinary language should decide against this all-inclusive classification. Private actors in war who drive trucks, clean latrines, cook meals, and perform other non-combat tasks are not typically recognized as mercenaries, though they *are* typically recognized as military contractors. Of course, one might argue that, while the term "mercenary" may indeed be unsuitable for support personnel, the term "military contractor" is equally unsuitable for those who engage in combat. But these hired killers are also under contract with belligerents. It seems that they may appropriately bear the moniker of "military contractor" *while also* bearing the moniker of "mercenary."[11] By contrast, military support personnel like truck drivers, custodians, and cooks do not easily bear the moniker of "mercenary" *while also* bearing the moniker of some further mercenary sub-class.

5.2 Weapons and the army

Having clarified the different private actors who animate Part II, let us turn now to an influential defense of wholesale military outsourcing. According to this defense, private armies, in which military contractors populate *all* positions on the chain of command, may be justifiable.[12] In such private armies, independent contractors may populate these positions, or PMSC employees may populate these positions, whether from a single PMSC or multiple PMSCs. The important feature of private armies is that they consist not only of private combat personnel and private support personnel (mechanics, medics, and so on) but also private commanders.

Fully private armies were the armies of choice for Thomas More's (2002 [1516]) Utopians. They also have historical precedent. They operated, for instance, in the Thirty Years' War (1618–48), where "almost all battles were fought completely by hired units" (Singer 2007: 29). Modern PMSCs like Blackwater might count as private armies if deployed alone to conflict zones. Indeed, J. Cofer Black, a former Vice President at Blackwater raised the possibility of PMSC deployment to Darfur (Weiner 2006): "We're low cost and fast.... The question is, who's going to let us play on their team?." By "team," of course, Black may have meant an army comprised of *both* regular military personnel and private contractors, but it does not require a great leap to envisage the deployment of a wholly private army. Michael Walzer (2008) did precisely that in his analysis of Darfur: "Whatever Backwater's motives, I won't join the 'moral giants' who would rather do nothing at all than send mercenaries"; while extensive military privatization is "mostly a bad idea," he argues, "private armies" are not "everywhere and always a bad idea." Deane-Peter Baker (2011: 161) has been even more explicit in endorsing "an intervention force made up *entirely* of contracted combatants" to carry out humanitarian missions (my emphasis added).

Perhaps the strongest defense of such private armies comes from Cécile Fabre, who argues that they are not just a permissible exception in humanitarian crises but are justifiable in any war so long as that war satisfies the just war criteria of just cause, proportionality, and non-combatant immunity. Fabre relies upon two fairly uncontroversial presuppositions to motivate her argument. First, states require weapons and armies for just defensive killings—namely for national defense, defense of others, and the enforcement of international norms. Second, in order to carry out just defensive killings, states are entitled to purchase weapons from private manufacturers (as they ubiquitously do). From these two presuppositions, Fabre (2012: 215–216) constructs the following argument by analogy:

> [P]olitical communities need the wherewithal to have acts of killing carried out in their name and with their authorization, in self-defence as well as in defence of others. And if a political community is at liberty to buy guns from private manufacturers for the aforementioned purposes—as it surely is—then it is also at liberty to buy soldiering services from those willing to provide them, irrespective of the latter's nationality. Moreover, if a political community has a right to pay for a standing army—as it surely does—then it also has a right to pay for a private army.

To be clear, there are two claims on offer in this argument. The first claim is that the purchase of guns from private entities is sufficiently analogous to the purchase of military services from private entities such that the justification for the former applies to the latter. Call this the *weapons-claim*. The second claim is that the purchase of private military services (and perhaps the formation of public armies) is sufficiently analogous to the formation of private armies such that the justification for the former applies to the latter. Call this the *army-claim*. The first claim posits that buying *some* private military services is justifiable, while the second posits that buying a *wide range* of private military services is justifiable.

In defending the weapons-claim, Fabre borrows from her analysis of mandatory rescue killings.[13] The position that she defends in that work begins by asserting the individual right to receive material resources for survival and then moves step-wise towards a purported right to purchase killing services (Fabre 2007). The progression may be reconstructed as follows.[14] *If* (*a*) one is entitled to receive food from a third party in order to avoid starvation, as one must be, then (*b*) one is entitled to receive a gun from a third party in order to defend oneself. In both cases, a material resource is needed for continued survival. But if one is entitled to receive a gun from a third party, then (*c*) one must also be entitled to receive assistance from a third party to thwart attack. The rationale that would permit (*b*) would also permit (*c*)—namely that material resources and services are both "fungible and scarce" and sometimes necessary "to pursue our ends" (Fabre 2007: 366). Finally, if (*d*) one is entitled to offer incentives like money to garner assistance—for instance, in the way that patients typically pay surgeons for operations (Fabre 2012: 220)—then one should be entitled to pay private killers to garner assistance. An affirmation of (*a*)–(*d*) entails approval of the weapon-claim: the purchase of guns from private entities is morally analogous to the purchase of military services and private killers may therefore be hired.

For now, I will assume *arguendo* that each step (*a*)–(*d*) is indeed justifiable and that the weapons-claim may be affirmed.[15] I will instead focus my criticism upon the army-claim. The army-claim, in some sense, builds upon the weapons-claim, though the relationship between the two is not fully specified. I take the army-claim to maintain that the purchase of private military services is comparable to the purchase of a private army in such a way that the *rationale* for the former applies to the latter—just as the rationale that we might offer for the employment of a public combat soldier may be used to ground the employment of a public standing army.[16]

The difficulty is that an argument for private killing simply does not suffice as an argument for the private performance of *all* military functions. It is true that the *raison d'être* of the military is killing. Thus, on first glance, the notion that a justification for private killing could serve as a justification for private armies should seem rather intuitive. But the fact that a state may permissibly outsource a single military role (killing) does not mean that it may outsource *any* military role.

Fabre has at least two strategies available to preserve the army-claim. One is to insist—in line with the weapons-claim—that killing and support (non-killing) services are *both* "fungible and scarce" and sometimes necessary to pursue our ends. Perhaps the blanket prohibition of certain killing services *or* support services would constitute an unjustifiable infringement of the individual right to self-defense and other-defense. If this were correct, then a move from the endorsement of private killers to an endorsement of private armies would be sound.

But the content of a right may be limited by competing values. To know whether competing values *do in fact* limit the right of self-defense, we must consider the full gamut of military roles, not just killing services. Militaries require interrogators, cooks, psychologists, lawyers, builders, engineers, drivers, chaplains, doctors, and so on. Indeed, the jobs that are essential to a military are jobs that are essential to almost any community of individuals. In that sense, the military functions as a microcosm of society—the very concerns that are aroused by outsourcing in civilian society may apply to the military. If any such functions must remain public, then the content of our right to self-defense will be limited to the enjoyment of military protection that is undertaken by armies with public personnel in those roles.

Given our concerns about privatization outside of the military, such functions may indeed abound. Section 1.5 indicated that some roles in civilian society, like garbage collection, are uncontroversial candidates for potential outsourcing, while others, like criminal adjudication, seem to be off limits. Something similar may be said of garbage collection and criminal adjudication in the military. Outsourcing garbage collection in a war zone may be uncontroversial, but populating military courts with judges who work for private companies would certainly raise concern. This simple point—that the privatization of some military functions, like criminal adjudication, are potentially problematic for the very reasons that the privatization of these functions in civilian society is problematic—forces us to look inside the military black-box and acknowledge that *many* values may be compromised by military privatization, just as we acknowledge this much in civilian society. The content of our right to self-defense and other-defense may thus be far narrower in a military context than Fabre supposes.

The second strategy available to Fabre, which she adopts in her later work on mercenarism, is to endorse *contingent* reasons for limiting the scope of the army-claim, thereby avoiding the unpalatable conclusion that *wholly* private armies are justifiable. Contingent objections, following Pattison (2014: 8–12), are those that cease to apply in the presence of optimal regulatory efficacy, while deeper objections are those that apply even in the presence of optimal regulatory efficacy. Fabre (2012: 238) considers the challenge that her view might entail a permission for governments to field exclusively private armies but maintains that she "is not convinced" that her view *does* entail such a permission. She gives three contingent reasons why the inclusion of some public, regular soldiers may be superior to their wholesale exclusion: the potential need for these soldiers to regulate private soldiers, their efficacy in combat, and the contributions of public soldiers to a diversified military (preventing undue dependence by governments on one type of force).

The problem with this second strategy is two-fold. First, it fails to recognize that there may be *deeper* reasons for objecting to private armies—in the same way that there may be deeper reasons for objecting to outsourcing in the civilian setting. Second, none of her explanations specify the set of functions that ought to be off limits to military outsourcing. At best then, her attempt to preserve the army-claim is incomplete, leaving open the hardest question that a theory of military privatization must confront: *which roles*, if any, ought not to be outsourced? At worst, her ad hoc solution reveals a deeper and perhaps inescapable weakness: the argument cannot provide deeper moral grounds for identifying these roles because of its singular focus on killing.

5.3 The governance argument

It is with an eye toward the broad spectrum of military services that we may now turn to the governance argument. According to the governance argument, the political power and decision-making authority of some military personnel in war are sufficient to demand that these functions not be outsourced. I refer to these individuals as *high-ranking military personnel.*

5.3.1 High-ranking military personnel

Recall from Section 1.7 that high-ranking military personnel are those who connect the military to the political system and the political system back to the military (Bland 1999: 15).[17] Typically, they might include flag grade officers (generals) and field grade officers (colonels and majors). But they are defined by the functions that they fulfill rather than the titles that they hold. These officers, I argue, exert political power over civilian decisions of military authorization and supply. Moreover, they make weighty strategic and tactical decisions *themselves* on behalf of a citizenry in battle.

To some, the definition that I offer of high-ranking military personnel will seem objectionably imprecise. There may be company grade officers and non-commissioned officers, for instance, who exert political power over civilian decisions. While this is certainly true, the argument of this chapter does not require specificity in identifying a list of high-level officers. We may reject A_3 according to the argument that I provide so long as we identify *some* military personnel whose positions ought not to be outsourced on the basis of their role in decision-making. Moreover, a theory of military supply is primarily valuable not because it enables us to identify *every* example of such functions—there will always be difficult cases—but because it provides a rationale for analyzing difficult cases.

As these considerations make clear, the governance argument is pitched towards public actors (e.g., states) who possess standing armies. Yet, as we have seen, non-governmental actors, like members of the Warsaw Ghetto uprising, might *publicly* authorize military force, thus potentially challenging the governance argument. Such public actors (uprisings, resistance movements, guerillas, and so on) will seemingly lack "officers," much less "high-ranking officers" of the sort that I have in mind. Still, the governance argument, I wish to suggest, may be applied to these more loosely organized, unconventional public actors. Even non-state entities that authorize military force may possess a command structure (of some sort) with certain contributors atop that command structure. Consequently, some individuals will function to connect the military forces to the political system and the political system back to the military forces. The governance argument insists that the responsibilities of *these* individuals must not be privatized—a formal system of officers and rank-and-file personnel is not necessary. For the sake of ease, I direct the governance argument at state-like actors that possess militaries with a formal command structure, but this is not meant to take away from the argument's wider applicability.

5.3.2 Public officials

The governance argument insists that high-ranking military officers serve as *public agents*. But what does it mean to serve as a public agent? These officers are not typically elected. They are appointed by government officials or promoted by those who have been appointed by government officials. It may not be clear, at this stage, what the governance argument aims to show by claiming that high-ranking military officers must be public agents.

In the analysis that follows, one is a public official (and not a private contractor) if, in one's governmental employment, one is constrained by a set of maximally precautionary institutional strictures to ensure the furtherance of collective interests. For instance, when a *public* high-ranking military officer is appointed to serve by a government official, the high-ranking military officer will be subject to a full menu of safeguards that her government employs to ensure proper representation: the military official who is appointed may receive a salary instead of a fee-for-service; she may serve at the pleasure of government officials and thus face potential dismissal at their discretion without fear of breaching contract; it may be that the appointed official is not personally liable (legally) for actions that are undertaken in her auspices as a governmental representative; she may receive a pension and health care; she may be eligible for awards and honors that are reserved for governmental employees; she may have to wear a special uniform. In short, governments aim to employ the full range of tools at their disposal to guarantee that the decisions of these officials match the interests of the collective.

Private agents, by contrast, are constrained by something less than such maximally precautionary strictures. Government officials may pay private agents a fee-for-service instead of a salary; they may hire individuals to serve according to the dictates of a contract rather than the pleasure of government officials; they may hold individuals who have been hired personally liable for their actions; they may not offer a pension or health care benefits; they may exclude such individuals from the awards and honors that are bestowed by the government; they may prohibit these hired individuals from wearing special uniforms. All such practices should be familiar manifestations of what, in everyday political discourse, we typically call "privatization." The important point is that private high-ranking military officers would be those who discharge the responsibilities of governance that I sketch in this chapter but who are held accountable by some set of constraints that are less robust than the constraints employed to bind public representatives.

Of course, this definition may seem ad hoc and philosophically wanting, particularly since governments that employ some of these practices and not others should not always be thought to *lack* publicly appointed employees. On some level, this criticism presents a challenge. This book is not in a position to offer an empirical analysis of which institutional practices *do* ensure that public officials—like individuals who now occupy the roles of high-ranking military officers in most states—serve collective interests and then argue that these are the institutional practices that must constrain high-ranking military officers. What I *do* suggest, however, is that high-ranking military officers must be public agents in the sense that they ought to be held to these maximal institutional strictures, whatever they may be, rather than the more minimal strictures that are associated with private contracts. This is what is meant when the governance argument concludes that high-ranking military officers must be public agents. They must be held to a set of institutional strictures that aims to maximally ensure sound representation.

5.3.3 Outsourcing decisions of authorization and continuation

The governance argument seeks to exploit a simple principle: our commitments to public legislative control over authorization, as developed in Part I, entail *certain* commitments to public military control over supply. Two propositions follow from an endorsement of exclusively public military authorization.[18]

The first proposition is that public officials who make decisions of military authorization may not outsource these decisions to the private sector. Recall from Section 3.4.2 that the risk-prevention argument does not require risk-bearers *themselves* to make decisions of military authorization but instead permits political representation. The current claim speaks to the question of whether public representatives may hire private agents to make decisions of war on their behalf; perhaps outsourcing these decisions would enable public officials to focus on other pressing political problems while ensuring that risk-bearers receive effective and efficient service when military authorization is needed. Call the public officials who delegate decision-making authority *public legislators* and the private officials who are granted such authority *private legislators*.[19]

Let us suppose that a great deal of legislative outsourcing is justifiable. Public legislators may hire private legislators to draft legislation, give advice, conduct polls, meet with voters, and so on.[20] In my view, even if a great deal of legislative outsourcing is justifiable, public legislators must ultimately retain a final say over the legislative decisions. Public legislators, in other words, must, at a minimum *oversee*, private legislators. By oversee, I mean that public legislators must manage the contracts of private legislators—by writing them, assuring compliance, terminating them when necessary—and must monitor their decisions so as to exercise final decision-making authority over the course of action that will be pursued. If, for instance, private legislators were to reach the decision that military force ought to be authorized, public legislators must review that decision and sign off on it (or not). The reason for such collective enforceable oversight is that risk-bearers have an obligation to exercise a final say over decisions of military authorization, and they cannot allow their representatives to turn over this final say to third parties.[21]

When public legislators *do* adequately oversee private legislators, then we ought to say that public legislators are the entities that authorize war; *they* are the ones who exercise a final say over such decisions, thereby relegating private legislators to advisory roles. Needless to say, the oversight must be substantial: as oversight becomes nominal and the "advice" of private legislators begins to dictate decisions of military authorization, then public legislators cease to retain exclusive authority over military authorization. But what I am suggesting here is that public legislators, who have been chosen to represent risk-bearers, must retain control over decisions of military authorization.[22]

A critic might demand a more satisfying explanation of what it *means* to be a public legislator. How does the representation of private legislators differ from the representation of public legislators? The answer is that public officials, as we saw in Section 5.3.2, are bound to the collective by a set of maximally constraining institutional strictures, while private officials, by definition, are bound by something less. In decisions of military authorization, the collective of risk-bearers has an obligation according to the risk-prevention argument to ensure that decisions of military authorization reflect the preferences of fellow risk-bearers. This, in turn, demands that members of the collective who do not themselves make (or directly contribute to) decisions of military authorization attempt to maximally constrain their chosen representatives—and only public legislators are so constrained.

The second proposition relevant to the governance argument that follows from an endorsement of exclusively public military authorization is that civilians who make decisions of military *prosecution*—for instance whether to continue fighting a war or not—must also be public legislators. Supposing that the first proposition is correct, an argument for the second proposition readily follows. Once a war is authorized, regular decisions are required to determine whether and how the war should be continued—and, at some point, a decision will be required to determine whether and how the war should be terminated. Call these *decisions of continuation*. It would seem that the requirement of public legislative control over decisions of authorization must also demand public legislative control over decisions of continuation. Like decisions of military authorization, decisions of continuation are decisions about whether (and how) the large-scale resort to arms for some cause is to be undertaken. Consequently, it would be difficult to insist that military authorization must be public but that decisions of continuation and termination may be *entirely* private. Any reason that we might propose to justify the former would apply to the latter. As a result, we ought to conclude that public legislators must retain ultimate control over at least *some* decisions of continuation.

5.3.4 The argument

Together, these propositions pave the way for two distinct defenses of the governance argument, which are argued in Sections 5.4 and 5.5: the defense from *political power* and the defense from *military decision-making*. Each defense is sufficient to show that high-ranking military personnel must be public agents. The two defenses are as follows:

Defense from Political Power:

P₁ If the decisions made by some set of individuals must be publicly discharged, then the responsibilities of those who exercise sufficient political power over such decisions must also be publicly discharged.

P₂ High-ranking military personnel exercise sufficient political power over public civilian decisions of military authorization and military supply for us to insist that the responsibilities of these individuals must be publicly discharged.

C_G High-ranking military personnel must be public agents.

Defense from Military Decision-Making:

P₃ If the decisions made by one set of individuals must be publicly discharged, then the decisions made by a second set of individuals must also be publicly discharged if the two sets of decisions are sufficiently similar in some morally decisive sense.

P₄ The decisions of high-ranking military personnel in military supply *are* sufficiently similar to the public decisions of civilians in military authorization in a morally decisive sense. The morally decisive sense in which they are similar is that both have the potential to cause extensive destruction on the scale that wars are fought.

C_G High-ranking military personnel must be public agents.

As we can see, the defense from political power and the defense from military decision-making both arrive at the conclusion of the governance argument (C_G). But they do so through appeal to different commitments.[23]

5.4 The defense from political power

Let us begin with the defense from political power. The first premise of this defense is a conditional whose antecedent is affirmed by the two propositions defended above (in Section 5.3.3): *if* the decisions made by some set of individuals must be publicly discharged, then the responsibilities of those who exercise sufficient political power over such decisions must also be publicly discharged (P₁). The two propositions showed that the responsibilities of civilian legislators who make decisions of military authorization and continuation must be publicly discharged. The question that we must then answer is whether high-ranking military officers exert political power over these civilian decisions to an extent that is sufficient to demand the public exercise of this power. But first, a defense of the connection between the antecedent and consequent of P_1 is needed. Why must those who exercise certain forms of power over public legislators be public agents?

My understanding of political power follows that of Robert Dahl (1957: 202–203): "A has power over B to the extent that he can get B to do something that B would not otherwise do." Thus, when I claim that high-ranking military personnel exercise political power over the decisions of civilian legislators, I mean that high-ranking military personnel are capable of getting civilian legislators to make decisions that they would otherwise not make. Of course, *many* actors may exercise power over public legislators, and yet we may not want to insist that all such individuals be public agents. For instance, the spouse of a public legislator may exercise power over her decisions, and yet we do not want to insist that the spouses of civilian legislators must be public agents. For this reason, I refer specifically to *political* power. My analysis is restricted to those agents who exercise power over legislators *in their capacity* as government employees or private contractors. The claim of P_1 is that individuals who exercise political power over a legislative decision must be public agents if the legislative decision itself must be publicly undertaken.

The plausibility of P_1 should be straightforward. We cannot simultaneously care about the public discharge of decisions while also being content if private actors are able to dictate these decisions. On any account of why legislative decisions ought to be public, private actors must not be permitted to co-opt the legislative process by substituting their judgments for the judgments of public legislators—to allow them to do so would just be to acknowledge that legislative decisions need not be public. This is precisely why we demanded that the advisory role of private legislators not be over-inflated. As the ability of private legislators begins to direct decisions of military authorization and continuation, the requirement of public military authorization/continuation seems to be violated. Surely then, we ought to conclude that *some* exercises of political power over public decisions are sufficient to demand that these exercises be public (the claim of P_1).

But our question is whether certain *military* contributions to civilian legislative decisions ought to be public or whether all such contributions may be private. Should high-ranking military personnel who participate in the shared enterprise of civil–military relations and thus exercise political power over decisions of military authorization and continuation be public officials? Or are we justified in restricting our attention to the civilian participants?

The second premise of the political power defense (P_2) insists that attention be paid *both* to civilian exercises of power and military exercises of power. More specifically, it claims that the exercise of political power by high-ranking military officials is indeed enough to demand that these officials be public agents. Such power is exercised both in civilian decisions of military authorization and civilian decisions of military continuation. Each will be considered in turn.[24]

5.4.1 Military power in civilian decisions of authorization

My contention is that high-ranking military personnel, like civilian legislators, have the power to drag states into wars that they should not fight and keep them out of wars that they should fight. In this sense, the political power of high-ranking military personnel in decisions of military authorization is substantial. Yet, traditional approaches to civil–military relations might argue otherwise. The prevailing presupposition from which explorations of civil–military relations begin is *civilian control*.[25] Civilian control, as traditionally understood, requires that civilians determine the policies of war, while soldiers implement the orders that follow. To be more precise, civilians in democratic regimes elect political leaders to represent their interests, and political leaders then issue judgments that are transmitted to high-ranking military personnel down the chain of command. To the extent that civilian control accords with this model, military authorization is *entirely* civilian-determined; military leaders are not involved in decisions of governance *ad bellum*. They simply execute the will of the civilian populace on the battlefield—the requirement of public military authorization has little bearing on the privatization of military personnel.

But few, if any, scholars of civil–military relations endorse this simplistic model. Civilian control "in the old sense," as it has been dubbed, is embraced neither as an empirical assessment of how militaries function nor as a normative prescription for how they should function.[26] The first problem that confronts the doctrine of civilian control in the old sense—two others will also be explored— is military influence over decisions of authorization. Civilians must rely upon military personnel for information and advice about war. If such input should not be privatized, then the function of high-ranking military leaders, who provide this input, should not be privatized. This, of course, is precisely what the first defense of P$_2$ maintains (in the political power defense of C$_G$).

The plausibility of this position becomes evident once we recognize that certain military contributions to civilian decisions of military authorization are *necessary* for such decisions to be justifiable (we may then readily see how these necessary contributions can be *decisive* in altering the course of military authorization one way or another). The need for military contributions to decisions of authorization stems first from the fact that military personnel are better positioned than civilian personnel—and are usually exclusively so positioned—to identify the threats that are capable of rendering decisions of military authorization just. As Walzer (2006: 74) reminds us, whenever war is considered, our initial questions must be "questions of fact, not of judgment": "Who started the shooting? Who sent troops across the border?" High-ranking military personnel are often the entities that are best equipped to answer these questions.

Furthermore, military personnel must instruct civilian leaders about the implications of war for military capability. Not all threats that permit the authorization of military force are sufficient to motivate authorization. Military personnel are uniquely capable of assessing concerns that transform permissible military authorizations into advisable military authorizations. In particular, high-ranking military officers are able to assess whether the military is ready for war, whether the deployment of forces will have a detrimental effect on the efficacy of other military efforts, whether readiness for future wars will be undermined by military authorization, whether a war is likely to pose long-term problems for troop morale and recruiting, and whether military action is capable of meeting the goals that a government wishes to achieve.

Most importantly, military personnel are positioned to assess the *jus ad bellum* criteria of probability of success and proportionality. In order for military authorization to be morally justifiable, the likelihood of success in military operations must be sufficiently high, and the harm inflicted upon enemy combatants must be proportionate to the motivating offense. Military officials, it would seem, are uniquely able to estimate the likelihood of success in military authorization. Likewise, though they should not alone judge whether a response is proportionate—as such a determination must ultimately hinge upon the values of civilian legislators—military officials are uniquely able to predict the potential harm that will be inflicted by military action, which is necessary for calculations of proportionality. The reasons for their unique positioning with respect to considerations of success and proportionality are threefold. First, civilian leaders typically lack the training that high-ranking military officers receive for approximating these probabilities and harms. If they do possess this training, say because they served as high-ranking military officers before entering politics, civilian leaders lack the time and resources to remain fully cognizant of the technology, budgetary concerns, mobilization patterns, troop levels and training that impinge upon the success and proportionality of war. Finally, even if civilian politicians are up to speed on these matters, military officers, because of their expertise and efficiency, must be entrusted to plan combat operations (subject, of course, to civilian approval for large operations). Since probability of success and proportionality are so intimately linked to the nuances of combat, it must be the planners who advise on the probability of success and proportionality of war. Only in this way will assiduous attention to detail be paid and proper respect for enemy combatants and non-combatants shown.[27] In light of these requisite military contributions to civilian decisions of military authorization—which derive from the civilian need for military assessments of threats, military capability, and the *jus ad bellum* requirements of success and proportionality—high-ranking military officers cannot be excised from the political process.

As I hope will be self-evident, these necessary contributions are capable of driving states into the wrong wars and keeping states from the right ones. High-ranking military personnel may ignore or overestimate threats; they may provide misinformation about the variables that transform a just war into an advisable war (troop-readiness, morale, and so on); and they may indicate that wars without a reasonable chance of success will succeed (or vice-versa) and may portray *dis*proportionate wars as proportionate (or vice-versa).[28] In short, though legislators possess the final say over military authorization, military officers control several empirical desiderata that help determine decisions of authorization. On occasion, these empirical desiderata may be *decisive*: information about the nature of threats, military capability, chances of success, and harms that will likely result *all* may tilt the scales of authorization one way or the other. Since high-ranking military personnel control such information, they are capable of supplanting their own judgments for the judgments of civilian legislators. In extreme circumstances, morality may even dictate that military officers refuse to mobilize troops (say, for a patently unjust war).

Given the power of high-ranking military personnel in decisions of military authorization, it does not seem that one can easily endorse the view that military authorization must be decided upon by public legislators while also maintaining the view that *all* military officers who exercise power over such decisions may be private agents. How could one care that the decision-makers in military authorization be public agents if one did not also care that a set of individuals who could overtake the decision-making process also be public agents? To do so, I think, would be to fetishize final-decision-making authority *ad bellum*. On this point, I agree with Yehuda Ben Meir (1995: 17), who writes that the issue for civil–military relations "is not who ultimately decides but rather the relative input of the armed forces and the civilian bureaucracy."

This point should be particularly compelling when we remember that no single legislator typically has the authority herself to authorize military force; she has the authority to cast a vote for military authorization.[29] While high-ranking military personnel do not cast a formal vote in this sense, they do (and must), like legislators, sign off on decisions of military authorization—confirming that the threat to which such a decision responds exists, that the military is ready to deploy, and that it will be successful and will not inflict harm above some magnitude that civilians have dubbed proportionate.

5.4.2 Military power in civilian decisions of continuation

We have now seen, according to the first case for P_2 in the political power defense of the governance argument, that the responsibilities discharged by high-ranking military personnel in decisions of military authorization must be publicly discharged. However, this view does not speak to military *supply*. Indeed, the military advisory role that is formally institutionalized by many governments— as the Joint Chiefs of Staff in the US and the Chiefs of Staff Committee in the UK—is often divorced from operational command. Members of the US Joint Chiefs of Staff, for instance, do not issue orders but rather stand outside the chain of command (United States Joint Chiefs of Staff 2015). It may be then that the high-ranking military officials who consult with civilian entities in authorizing war must be public agents but that the military personnel who participate in the supply of war need not be public agents.

My second defense of P_2 insists that the political power exercised by high-ranking military personnel in civilian decisions of military supply is also sufficient to demand that these personnel be public agents. We saw above that civilian legislators must control decisions of continuation just as they must control decisions of authorization. Yet, once we acknowledge the need for public control over decisions of continuation, constraints over military contributions to these decisions follow in just the same way that they did in the previous sub-section. There, we identified a demand for public military contributions to public legislative decisions of authorization that followed from a demand for public legislative decisions of authorization. Here, we identify a demand for public military contributions to public legislative decisions of continuation that follow from a demand for public legislative decisions of continuation. My position is that high-ranking military personnel, like civilian legislators, have the power to keep states in wars that they should exit and exit wars in which they should remain. For that reason, the political power defense of the governance argument must be affirmed as a limitation on private military supply.

The political power of high-ranking military officers over civilian decisions *in bello* gains its force from the need for civilians to rely upon military personnel to provide ongoing and accurate empirical assessments of how a war is proceeding. Specifically, military officers in the field must provide civilians with information about whether military objectives are being fulfilled, whether the fulfillment of these objectives is being achieved with ease or difficulty, and whether further objectives may be fulfilled given the material resources and troops that are available. Military officers, in other words, must describe outcomes *in bello* and provide feedback to civilian personnel that will enable informed decisions about continuation.

These contributions underscore the second problem with civilian control "in the old sense" that was alluded to above. Whereas the first problem focused upon the inability of civilians to make decisions of authorization in the absence of military input, the second problem focuses upon the inability of civilians to conduct a war in the absence of military input. Without military input, civilians simply have no sound basis upon which to continue or terminate a war. The first two problems of civilian control "in the old sense" together highlight the role of military personnel in shaping civilian policy decisions and hence constructing the orders that flow from civilian leaders down the chain of command. Military personnel do not merely take orders from civilians; they inescapably contribute to the orders that are dispensed.

But are the necessary contributions of high-ranking military personnel to decisions of continuation as controlling as their contributions to decisions of military authorization? As I hope is clear, they are certainly capable of keeping states in wars that they should exit and driving them out of wars in which they should remain. Through misrepresentation of "facts on the ground," high-ranking military officers can shift civilian policy to suit their commitments. Under certain circumstances, they may be obligated to refuse to go along with civilian policy *in bello*. Marybeth Ulrich and Martin Cook (2006: 171) write about those instances when military leaders believe that the goals of a government are "unattainable by use of the military instrument of national power": "Even though their subordination clearly requires execution of legal orders, surely there is some ethical obligation of the most senior military leaders not to soldier on in pursuit of a policy they sincerely believe to be doomed to failure."

Through both a control over the empirical desiderata that are necessary for the continuation (or termination) of war and through the capacity of high-ranking military officers to refuse to carry out such policies (and instead carry out the policies that they believe are alone morally defensible), these officers exercise substantial control over the direction of war.

In some ways, the contributions of high-ranking military personnel to civilian decisions in military supply may be more significant than their contributions to military authorization. This is because civilian leaders have long showed a deference to military expertise *in bello* that seems to go beyond the deference that is showed *ad bellum*. Consider a conversation that President Bush had with high-ranking military officials after the invasion of Iraq. As Bush reported, he asked his generals whether they had the right tactical plan and appropriate troop levels.

> And they looked me in the eye and said, yes, sir, Mr. President. Of course, I listened to our generals. That's what a President does. A President sets the strategy of war and relies upon good military people to execute that strategy.
>
> (in Ulrich and Cook 2006: 173)

Given the troubled US invasion and occupation of Iraq, this quote serves to remind us how civilian officials must not grant military officers *too much* discretion *in bello*. Civilians must exercise the final say over troop levels and over many other questions of continuation and termination. But the quote also reminds us how much power high-ranking military officers may exert over civilian policy *in bello*.

In light of this power, our insistence that civilian decisions *in bello* must be public should compel us to insist that certain military contributions to these decisions—namely those provided by high-ranking military personnel—should also be public. Any motivation that one might have for demanding public civilian control over decisions of continuation should also demand the public discharge of *some* military contributions to these decisions. With this in mind, the second defense of P_2, which is the defense that concerns military supply, is complete: high-ranking military officers *do* seem to exercise sufficient political power over public civilian decisions of military authorization and military supply for us to insist that the responsibilities of these officers must be publicly discharged. The conclusion of the governance argument (C_G)—that high-ranking military officers must be public agents—has now been demonstrated via one strategy: the defense from political power, according to both *ad bellum* and *in bello* considerations.

Before moving to a second defense of the governance argument, one objection should be considered. The objection is that the defense from political power has claimed too much. After all, state department officials, civilian intelligence agents, and many others may exercise the power to keep us out of the right wars and drive us into the wrong wars. Indeed, private technical experts and scholarly experts (in law, ethics, strategy, etc.) may exercise so much influence over civilian legislators that they too have the power to dictate decisions of military authorization.

But the governance argument would gladly embrace each of these observations. The power of state department officials and civilian intelligence agents may simply provide a good reason why *they too* must be public agents. Alternatively, perhaps the power of private experts may provide good reason why, in contrast to state department officials and civilian intelligence agents, their influence over military authorization should be limited. The governance argument, in other words, may generate a set of entailments that goes beyond high-ranking military officers to include civilians outside of the legislature who are intimately involved in the governance of war. This much should only strengthen the governance argument by widening its application to other plausibly core governmental functions.

5.5 The defense from military decision-making

We have now seen that, because high-ranking military officers exercise such great political power over civilian decisions of military authorization, they too, like civilian legislators, must be public agents. I now wish to consider another independent defense of the governance argument: the defense from military decision-making. This defense concerns not the contributions of high-ranking military personnel to civilian decisions but the decisions that these high-ranking military personnel *themselves* make in the field. Recall the first premise of this defense (P_3): if the decisions made by one set of individuals must be publicly discharged, then the decisions made by a second set of individuals must also be publicly discharged if the two sets of decisions are similar in a morally relevant sense. In the first subsection below, I will motivate support for this premise by providing a conceptual framework that elucidates how two decisions might be similar to one another in a morally relevant sense. I then seek to show that the decisions of high-ranking military personnel in military supply *are* similar to the decisions of civilians in military authorization in a morally relevant sense (P_4).

5.5.1 Legislative and military decisions

Civilian decisions of military authorization (D_A) and military decisions of supply (D_S) may be similar to one another in at least two morally relevant senses. According to the first morally relevant sense, the *specific normative principles* that are developed to govern one may govern the other. Consider the theory of D_A that I offered in Part I: private military authorization imposes grave risks upon others (*risk-imposition of war premise*) and, when such grave risks are imposed, decisions must be withdrawn from the private sector and reserved for public actors (*all affected fundamental interests premise*); these public actors, in turn, have a further responsibility to protect individuals who have been disarmed (*compensation argument*). This theory of exclusively public military authorization, like any such theory, is constructed from facts (F) and principles (P). The risk-imposition of war premise is a fact, while the all affected fundamental interests premise and the compensation argument are principles.[30]

While the principles of the risk-prevention and compensation arguments were developed to theorize about D_A, the principles may nonetheless remain apposite for theories of D_S. Suppose, for instance, that high-ranking military personnel were charged with decisions that imposed grave risks upon others. Depending upon the scale of these risks, the all affected fundamental interests premise may be applied so that decisions of military supply must be removed from the private sector and reserved for public discharge. A principle that accompanies one set of facts pertaining to civilian decisions of authorization, in other words, is potentially serviceable for a new set of facts pertaining to military decisions of supply. These considerations underscore one morally relevant sense in which civilian decisions of authorization may be similar to military decisions in supply: the specific normative principles that govern D_A may be activated by the facts of military supply to generate a requirement about D_S. Call this a *particularized similarity*.

But there is a second potential morally relevant similarity between D_A and D_S—one that is independent of the specific theory on offer and, by extension, is independent of the principles and facts underpinning that theory. Imagine that several theories of exclusively public D_A are defensible. All of these theories conclude that private military authorization is unjustifiable and that D_A must be withdrawn from the private sector and reserved for public discharge. In other words, one set of principles (P_X), corresponding with a set of facts (F_X), requires that we withdraw D_A from the private sector and reserve it for public discharge; but other theories of authorization—including a second theory derived from P_Y and F_Y, a third derived from P_Z and F_Z, and so on—also require that we withdraw D_A from the private sector.

Without reference to any of these particular theories, it may be that a requirement for the public discharge of D_A is also applicable to D_S. The reason is that the decisions *themselves* may be sufficiently similar so that any theory of D_A would apply to D_S. The supply of war, in other words, may consist of mini-authorizations (of sorts)—for instance if generals make decisions in supply that look just like decisions of authorization. Whether our insistence upon the public discharge of D_A is derived from P_X and F_X, P_Y and F_Y, or P_Z and F_Z is unimportant; our *conclusion* about D_A—as opposed to our specific *rationale* for the derivation of any such conclusion about D_A—may generate conclusions about D_S. Call this a *generalized similarity*. The military decision-making defense of the governance argument relies primarily upon generalized similarities for ecumenical reasons. But I also point out how particularized similarities are capable of grounding the argument.

5.5.2 *Military decision-making* in bello

Having motivated support for P_3 of the military decision-making defense, let us turn to P_4. P_4 claims that the decisions of high-ranking military personnel in military supply are sufficiently similar to decisions of civilian legislators in military authorization that a commitment to the public performance of the latter entails a commitment to the public performance of the former.

This premise—P_4—speaks to the final problem of civilian control "in the old sense." An understanding of why should begin to motivate support for the plausibility of this premise. The preferences of civilians cannot always be seamlessly transmitted from citizens to soldiers down the chain of command. Citizens must sometimes rely upon political leaders in government and military leaders on the battlefield to *interpret* these preferences and make decisions accordingly. Thus, the battalion commander who is ordered to withdraw his troops from a confrontation is not acting only on collective preferences that have been *directly* transmitted by the civilian populace (or even by civilian legislators) down the chain of command. He is also acting on the orders of military officers who have themselves determined that a battalion-retreat would best serve the interests of the collective. High-ranking military officers who give these orders may receive civilian input. But they must often make decisions on behalf of their collective about how attacks will proceed, which cities should be occupied, how many men and women ought to be deployed, and, more generally, what tactical objectives should be pursued.

The decisions of military officers on the battlefield differ from their policy contributions in two fundamental ways. First, military officers must often use their *sole discretion* when making decisions on behalf of the collective. By contrast, in policy deliberations, military officers provide information and advice to political leaders who then make final decisions. Of course, the decisions of military officers *in bello* are always made within the constraints of civilian-determined policies of engagement. But, on the battlefield, both because quick decisions are often required and because communication between civilians and military personnel is sometimes impossible, military leaders must assume a more complete responsibility for these decisions than for policy decisions.

The second critical point to note regarding military decision-making is that, while military officers must exercise great discretion in decisions of military supply, these decisions *typically* have far less potential for destruction than decisions of military authorization. Thus, when we speak of the sub-authorizations of military commanders, we must not overstate the similarities to most military authorizations.

Yet, it seems that if we are committed to public military authorization, we must be committed to the public discharge of *certain* decisions in the field. The reason—according to a generalized account—is simply that wars vary tremendously in scale, and sub-authorizations by military commanders may be larger than the very authorization of war in the first place. A small war might see the deployment of 5,000 soldiers upon authorization. But, in larger wars, where societies authorize the deployment of, say, 200,000 soldiers, many high-ranking military personnel command groups of soldiers that exceed 5,000 in number. This is not to claim that there are no differences between a state that authorizes an invasion, say, with 5,000 soldiers and a colonel who commands 5,000 soldiers to take a city. There are many differences. But it would seem misguided to insist that legislators who make decisions of military authorization must always be public agents but high-ranking military officers who make decisions in the field need *never* be public agents when the decisions made by the latter some-times entail *more* destruction than the decisions made by the former.

With this generalized constraint in mind, we might also observe how the par-ticularized constraints imposed by the risk-prevention argument apply: high-ranking military commanders have the potential to escalate the risks imposed in war by a level of magnitude that compares to the risks that are imposed by the authorization of small-scale wars. Put differently, the risks that high-ranking military officers might themselves impose upon direct and indirect-risk-bearers through their decisions on the battlefield surely may outpace the risks of *some* private military authorizations. This point should be particularly compelling when we remember that, unlike private actors, military personnel will typically possess access to sophisticated and extremely destructive technology that private actors often lack.

In light of both generalized and particularized constraints, we ought to affirm P_4 of the defense from military decision-making. The decisions of high-ranking military officers in military supply *are* similar to the public decisions of civilians in military authorization in a morally decisive sense. The morally decisive sense in which they are similar is evidenced both by generalized and particularized considerations. The governance argument must therefore be embraced according to the defense from military decision-making: high-ranking military officers must be public agents.

None of this is to deny the critical need for a chain of command to ensure that collective preferences guide decisions on the battlefield. I will argue, in Chapter 8, that authorizing entities *must* place all risk-imposing agents of war, including con-tractors, under a unified chain of command precisely so that collective preferences guide decisions *in bello*. The point of my defense from military decision-making is that high-ranking military officers fulfill a similar role to legislators—*both* must seek to understand and act upon collective preferences. For that reason, both must be public agents. Those who serve beneath high-ranking military officers on the chain of command do not carry out these governance functions and therefore need not be public officials on account of the argument presented in this chapter.

5.6 Conclusion

Defenses of private armies, as endorsed by Fabre and others, fail to recognize that militaries are heterogeneous collections of individuals who fill both combat *and* non-combat roles.[31] The military must not be mistaken for a mere collection of killers. Rather, the military must be understood as an assortment of indi-viduals with a wide range of functions who contribute to the prosecution of war. Killing, in other words, must not be the only concern for scholars who theorize about military privatization. The very principles that push us to reject privatiza-tion in certain spheres outside of the military should push us to reject privatiza-tion in certain spheres within the military.

This chapter has examined one such sphere: governance. I have provided two defenses for the conclusion that decision-making responsibilities in war (i.e., those responsibilities exercised by high-ranking military personnel) must not be privatized. First, high-ranking military personnel exercise considerable political power over civilian decisions of military authorization and military supply. Second, high-ranking military personnel make decisions in military supply that are similar to decisions of military authorization in a morally relevant way. According to both defenses, the responsibilities that are discharged by high-ranking military personnel must be publicly discharged. The end result is that private armies may be rejected, and A_3 may be foreclosed as a justifiable arrangement for the provision of military force.

Notes

1 See Section 1.3. These two functions may be carried out by the same entity or by different entities.
2 A "publicly controlled military" is just a military that prosecutes wars for a public actor.
3 It might also be that, in a partially outsourced public monopoly over force (A_4), publicly controlled militaries are composed of public personnel in some wars but private personnel in others.
4 Note that Fabre (2012: 208–238) is largely similar to Fabre (2010). However, she adds one crucial paragraph of caveats to the conclusion of her later work (2012: 238). I will examine these caveats in due course.
5 For a critical view of the terminology that I employ, see Coady (2008: 223). Coady believes that the "term 'contractors' both symbolizes and adds to the confusion"; the confusion, according to Coady, stems from the "bewildering variety of tasks" that contractors perform. But it is precisely this bewildering array of tasks that makes the term "contractors" helpful. I am interested in the full range of private military actors in war.
6 Note that Fabre applies this definition to mercenaries. But I believe that the definition effectively captures common understandings of the term "military contractor." More will be said about this below.
7 Throughout this book, I have referred to military protection as a *good* that societies provide to members according to one of the arrangements A_0–A_5. In the context of a society providing some good g to its members—whether g is military protection, health care, education, etc.—I was careful to note that "the provision of goods" included the provision of both materials and services. See Section 1.3. These considerations should not be neglected. "Goods" refer to both "materials and services." Whenever the term "military contractor" is used, however, I am referring specifically to those private agents who supply services rather than materials. In other words, military contractors are those entities that participate in the provision of some good g, namely military force, through the contribution of their services.
8 Private entities that provide *both* materials and services to public entities, like weapons manufacturers who maintain and operate missile systems that they sell, will be referred to as military contractors. Providing military services is necessary and sufficient for classification as a military contractor.
9 Labeling such enterprises has become the subject of much scholarly debate. As explained in Section 1.5, I follow Pattison in using the term "PMSC" rather than the terms "private military company (PMC)" or "private security company (PSC)." But I do not follow Pattison's definition of PMSCs. Pattison (2014: 17) defines PMSCs as "private firms that provide military and/or security services that involve or assist the use of force beyond the borders of their own or their client's political community."

One problem with this definition is the emphasis on force used *outside* of political communities (whether the contractor's political community or that of the authorizing entity). On Pattison's definition, Blackwater would not count as a PMSC if it were hired by the US government to fight a war on US soil in response to enemy invasion. This strikes me as counterintuitive.

10 If companies that do not seek *profit* should be thought somehow to stand outside of deliberations on military privatization, we must remember that military contractors, by definition, receive *some* payment, if not a lucrative payment. These not-for-profit companies therefore remain within our purview. By contrast, those entities that receive *no* payment are taken to be *volunteers* (not military contractors) and are beyond the purview of our investigation into privatization.

11 This analysis purposefully leaves aside the fraught question of what constitutes a mercenary. On my account, a mercenary is just one subtype of contractor. Beyond that, I am agnostic on whether mercenaries must, by definition, be involved in combat or carry foreign passports or possess pecuniary motives. These features are all considered in Part II—but whether any are *necessary* to qualify an individual as a mercenary is beside the point of my analysis. For important contributions to this definitional debate, see Singer (2007: 40–48), Percy (2007: 12–14), Steinhoff (2008: 17–29), Coady (2008: 205–211), Krahmann (2010: 5–8), Baker (2011: 32–35), Fabre (2012: 209–210), and Pattison (2014: 19–20).

12 Throughout this book, the term "army" is used synonymously with the term "military." As such, private armies may include individuals who not only perform military operations on land but also at sea and in the air. We must be careful to note that private militaries may be publicly controlled *or* privately controlled. However, unless otherwise specified, when I use the term private army, I mean to connote a publicly controlled private army.

13 Fabre (2012: 215, fn 8) writes that "giving a gun to someone who needs it in self-defence and killing (for free) that person's attacker are relevantly analogous. The argument applies, *mutatis mutandis*, to the act of selling a gun and the act of killing against payment."

14 A reconstruction is needed, because the theory that Fabre defends in the context of mandatory rescue killings concerns the *sale* (or *donation*) of weapons rather than the *purchase* (or *receipt*) of weapons. This subtle difference is not particularly important here and will therefore not be addressed.

15 Perhaps the most vulnerable place in Fabre's argument is the move from (*b*) to (*c*). I consider this move in Chapter 8.

16 One might think that the army-claim is making a far more innocuous point—which is that the private military personnel who serve a state need not be hired specifically for a single war or part of a single war. Just as a state may employ public military personnel during peacetime, so too may the state hire private personnel during peacetime—to train them, preserve readiness, ensure availability, and so on. I certainly believe that this more innocuous claim is correct. If a state is justified in hiring some set of private contractors to fight in war, then it would seemingly be justified in hiring *that set* of contractors before war begins and after war has ended. But this does not mean that a state could hire a full private army of contractors; it may still be that some military services are off limits to contractors—whether the contractors are hired to perform these services before, during, or after a war. I take Fabre to mean that *no* military services are off limits to contractors. She writes that "there is nothing inherently objectionable about mercenarism" and defines mercenarism to include not just killing in war but participation broadly speaking (2010: 539–540). I therefore do not think that the interpretation of the army-claim under examination in this footnote is the correct one.

17 This was how Colin Powell described high-ranking military officers. He was originally quoted in Woodward (1992: 154).

18 When I refer to "exclusively public military authorization," I am referring to the combination of public monopolization over military authorization and public military authorization.

19 I use the term "legislator" to refer to those officials who make decisions of military authorization on behalf of a collective. The term is somewhat of a misnomer since these officials may include members of the executive branch as well. But, for the sake of stylistic ease, I simply take legislators to be the representatives of citizens who make decisions of military authorization on their behalf. Further nuance about the relationship between the legislature and the executive is not explored.

20 We should observe that political officials today seem to regularly outsource legislation in this manner, allowing lobbyists and special interest groups to exert influence over lawmaking and even draft provisions of legislation.

21 Political representatives typically exercise decision-making authority in modern democracies. In theory, of course, the members of a democracy could also retain control by reaching decisions through referenda.

22 When neither the collective nor its representatives are capable of making (or participating in) such decisions, as we saw in cases of emergency authorization (Section 4.6), then third parties may rely upon secondary indicators of consent to authorize war on their behalf.

23 This book does not probe questions of representation. If the governance argument is correct, a perfectly reasonable question to ask is in what sense must high-ranking military personnel represent the interests of those whom they serve? It is the same question that must be asked about public legislators. What I urge is simply that favored theories of political representation, as developed for legislators, be applied to military personnel. There are, of course, many differences between the two, the most important of which is that legislators are often elected and high-ranking military officers are not. Consequently, certain modes of political representation that may be particularly applicable to legislators may not be so applicable to military officers. But, if my argument is correct and high-ranking military officers *do* discharge responsibilities that are similar in some morally relevant respect to the responsibilities that are discharged by legislators, then these military officers may be asked to represent citizens in the way that legislators are asked to represent citizens (to an extent). For literature on legislative representation, see Pitkin (1967), Beitz (1989: ch. 6), Dahl (1989), Urbinati (2000), Young (2000), Mansbridge (2003, 2009), Hardin (2004), Saward (2009), and Pettit (2010). See Dovi (2014) for a helpful review.

24 A caveat must be acknowledged here. Even if I am correct about the power exercised by high-ranking military personnel in decisions of military authorization (the first set of considerations), *still* military supply may remain wholly private while public agents serve as high-ranking military officers. Only the second set of considerations, which concerns the exercise of military power in civilian decisions of continuation, speaks to military *supply*. The first set of considerations simply sharpens my accounts of military authorization and monopolization (as defended in Part I). But because this added depth is crucial for a complete understanding of military contributions to war—and thus for a theory of military privatization—I will provide an elaboration. Such an elaboration appears in the current chapter rather than Part I only because of its affinity to the second set of considerations that motivates the political power defense of the governance argument. I trust that the benefit of this approach will become clear as the discussion unfolds.

25 For helpful texts in civil–military relations upon which the governance argument relies, see Huntington (1959), Janowitz (1960), Bland (1999), Feaver (1996, 1998, 1999, 2003), Burk (2002), Cook (2002–3).

26 The term "civilian control in the old sense" is quoted in Ben Meir (1995: 10). See Lovell (1974: 13).

27 For a similar point, see Cook (2002–3: 26): "among the moral tests use-of-force decisions must meet is that there be a reasonable hope of success," which means "bringing
about the desired result with an acceptably proportionate amount of destruction.
Professional military officers possess expertise in judging the capabilities of the
military instrument of power."
28 On the point that PMSCs are already beginning to exert power in this way, see Pattison (2014: 91–92) and Kinsey (2009: 104).
29 In some states, of course, like the US, the executive branch is permitted to initiate war
without a legislative vote. Typically, however, such permission is limited in temporal
scope. More to the point, I have argued that decisions of military authorization must
be collective—for that reason, these decisions *ought* to be made by a vote of public
legislators.
30 My discussion here draws on Cohen (2003), but the substance of my point is not
contingent upon his. Cohen argues that some principles are fact-independent. Whether
the principles that I identify here are fact-independent or fact-sensitive is tangential to
my discussion.
31 Note that Fabre herself would likely reject the governance argument since she does not
subscribe to the just war requirement of legitimate authority. See Fabre (2008; 2012:
141–156). See also Sections 2.3–2.5 for more on Fabre's account of legitimate authority.

References

Baker, Deane-Peter (2011). *Just Warriors, Inc.: The Ethics of Privatized Force* (London:
Continuum International Publishing).

Beitz, Charles (1989). *Political Equality: An Essay in Democratic Theory* (Princeton:
Princeton University Press).

Ben Meir, Yehuda (1995). *Civil Military Relations in Israel* (New York: Columbia University Press).

Bland, Douglas (1999). "A Unified Theory of Civil–Military Relations," *Armed Forced
& Society*, Vol. 26, No. 1, pp. 7–25.

Burk, James (2002). "Theories of Democratic Civil–Military Relations," *Armed Forces &
Society*, Vol. 29, No. 1, pp. 7–29.

Coady, C.A.J. (2008). *Morality and Political Violence* (Cambridge: Cambridge University Press).

Cohen, G.A. (2003). "Facts and Principles," *Philosophy and Public Affairs*, Vol. 31, No.
3, pp. 211–245.

Cook, Martin (2002–3). "The Proper Role of Professional Military Advice in Contemporary Uses of Force," *Parameters*, Vol. 33, No. 4, pp. 21–33.

Dahl, Robert (1957). "The Concept of Power," *Behavioral Science*, Vol. 2, No. 3,
pp. 201–215.

Dahl, Robert (1989). *Democracy and Its Critics* (New Haven: Yale University Press).

Dovi, Suzanne (2014). "Political Representation," in Edward N. Zalta (ed), *The Stanford
Encyclopedia of Philosophy* (Spring 2014 Edition). Available at http://plato.stanford.
edu/entries/political-representation/ (accessed December 28, 2015).

Fabre, Cécile (2007). "Mandatory Rescue Killings," *The Journal of Political Philosophy*,
Vol. 15, No. 4, pp. 363–384.

Fabre, Cécile (2008). "Cosmopolitanism, Just War Theory, and Legitimate Authority,"
International Affairs, Vol. 84, No. 5, pp. 963–976.

Fabre, Cécile (2010). "In Defence of Mercenarism," *British Journal of Political Science*,
Vol. 40, No. 3, pp. 539–559.

Fabre, Cécile (2012). *Cosmopolitan War* (Oxford: Oxford University Press).

Feaver, Peter (1996). "The Civil–Military Problematique: Huntington, Janowitz, and the Question of Civilian Control," *Armed Forces & Society*, Vol. 23, No. 2, pp. 149–178.

Feaver, Peter (1998). "Crises as Shirking: An Agency Theory Explanation of the Souring of American Civil–Military Relations," *Armed Forces & Society*, Vol. 24, No. 3, pp. 407–434.

Feaver, Peter (1999). "Civil–Military Relations," *American Review of Political Science*, Vol. 2, pp. 211–241.

Feaver, Peter (2003). *Armed Servants: Agency, Oversight, and Civil–Military Relations* (Cambridge: Harvard University Press).

Hardin, Russell (2004). "Representing Ignorance," *Social Philosophy and Policy*, Vol. 21, No. 1, pp. 76–99.

Huntington, Samuel (1959). *The Soldier and the State: The Theory and Politics of Civil–Military Relations* (New York: Vintage Books).

Janowitz, Morris (1960). *The Professional Soldier: A Social and Political Portrait* (New York: The Free Press).

Kinsey, Christopher (2009). *Private Contractors and the Reconstruction of Iraq* (Oxford: Routledge).

Krahmann, Elke (2010). *States, Citizens, and the Privatization of Security* (Cambridge: Cambridge University Press).

Lovell, John (1974). "Civil–Military Relations: Traditional and Modern Concept Reappraisal," in Charles L. Cochran (ed), *Civil–Military Relations: Changing Concepts in the Seventies* (New York: Free Press).

Mansbridge, Jane (2003). "Rethinking Representation," *American Political Science Review*, Vol. 97, No. 4, pp. 515–528.

Mansbridge, Jane (2009). "A Selection Model of Representation," *The Journal of Political Philosophy*, Vol. 17, No. 4, pp. 369–398.

More, Thomas (2002 [1516]). *Utopia*, edited by George M. Logan and Robert M. Adams (Cambridge: Cambridge University Press).

Pattison, James (2014). *The Morality of Private War* (Oxford: Oxford University Press).

Percy, Sarah (2007). "Morality and Regulation," in Simon Chesterman and Chia Lehnardt (eds), *From Mercenaries to Market: The Rise and Regulation of Private Military Companies* (Oxford: Oxford University Press).

Pettit, Philip (2010). "Representation, Responsive and Indicative," *Constellations*, Vol. 17, No. 3, pp. 426–434.

Pitkin, Hannah Fenichel (1967). *The Concept of Representation* (Berkeley: University of California Press).

Ryan, Cheyney (2009). *The Chickenhawk Syndrome: War, Sacrifice, and Personal Responsibility* (Lanham: Rowman and Littlefield).

Saward, Michael (2009). "Authorisation and Authenticity: Representation and the Unelected," *The Journal of Political Philosophy*, Vol. 17, No. 1, pp. 1–22.

Singer, Peter Warren (2007). *Corporate Warriors: The Rise of the Privatized Military Industry*, updated edition (Ithaca: Cornell University Press).

Steinhoff, Uwe (2008). "What Are Mercenaries," in Andrew Alexandra, Deane-Peter Baker, and Marina Caparini (eds), *Private Military and Security Companies: Ethics, Policies, and Civil–Military Relations* (New York: Routledge), pp. 19–29.

Ulrich, Marybeth P. and Martin L. Cook (2006). "US Civil–Military Relations Since 9/11: Issues in Ethics and Policy Development," *Journal of Military Ethics*, Vol. 5, No. 3, pp. 161–182.

United States Joint Chiefs of Staff (2015). "About the Joint Chiefs of Staff." Available at www.jcs.mil/About.aspx (accessed December 28, 2015).

Urbinati, Nadia (2000). "Representation as Advocacy: A Study of Democratic Deliberation," *Political Theory*, Vol. 28, No. 6, pp. 758–786.

Walzer, Michael (2006). *Just and Unjust Wars: A Moral Argument with Historical Illustrations*, 4th edition (New York: Basic Books).

Walzer, Michael (2008). "Mercenary Impulse: Is there an Ethic that Justifies Blackwater?," *The New Republic*, March 12, pp. 20–21.

Weiner, Rebecca Ulam (2006). "Peace Corp." *The Boston Globe*, 23 April.

Woodward, Bob (1992). *The Commanders* (New York: Simon and Schuster).

Young, Iris Marion (2000). *Inclusion and Democracy* (Oxford: Oxford University Press).

6 Punishment

The previous chapter argued that public, rather than private, military personnel must populate positions in the officer corps atop the chain of command. But, as it stands, my account of public military supply is incomplete. We must ask whether there are other military positions besides the positions of high-ranking military officers that ought to be reserved for public performance. The vast majority of military personnel are charged with decision-making responsibilities that are insufficiently weighty to invoke the governance argument of Chapter 5. If the only limitations imposed upon military outsourcing derive from the governance argument, then a state would be free to privatize the functions carried out by most of its officers and all of its rank-and-file personnel. This chapter presents an argument, dubbed the *punishment argument*, that aims to further curtail the scope of permissible military outsourcing. According to the punishment argument, militaries require some set of individuals to exercise command, and those who do so must be public agents. The reason is that commanders are required to issue intrusive forms of punishment, such as imprisonment, and these intrusive forms of punishment should only be dispensed by public actors.

In defending the punishment argument, I define "command" according to US military law.[1] To exercise command over a unit is to "be the individual chiefly responsible for maintaining discipline in that organization" (United States Army 2011: 5).[2] Commanders are tasked with "properly training their Soldiers" and remaining "vigilant in inspecting the conduct of all persons who are placed under their command" (United States Army 2014: 2).[3] The discipline, training, and supervision provided by commanders help guide the performance of soldiers on the battlefield.

The punishment argument begins from a fairly uncontroversial proposition: collective authorizing entities must rely on individuals to exercise command over at least *some* individuals in war. Just as collective authorizing entities require high-ranking military officers to connect the political system to the military (as we saw in the previous chapter), they require commanders to connect high-ranking military officers to the rank-and-file.[4] The reason for this requirement is two-fold. First, as the risk-prevention argument insists, collectives must not impose substantial risks upon the fundamental interests of others without consultation, and commanders are needed to ensure that rank-and-file personnel impose only collectively *chosen* risks. A collective that authorizes war and takes *no* steps to ensure that its military imposes the chosen distribution of risks (rather than an alternative distribution) would show improper respect for the fundamental interests of risk-bearers. Call this the *representative rationale* for military command. The second justification for military command is that authorizing entities have a duty to promote compliance with the requirements of *jus in bello.*[5] Commanders are needed to ensure that rank-and-file personnel do not wrongfully target non-combatants or inflict disproportionate damage on combatants. Call this the *jus in bello rationale* for military command. To achieve representative and morally permissible behavior on the battlefield, collectives must assign commanders to discipline, train, and supervise some (if not all) rank-and-file personnel.

The question is whether these commanders may be private agents. This chapter argues that they may not. It proceeds as follows. First, in order to contextualize and introduce the punishment argument, I briefly consider how current military practices have exposed a gaping deficiency in military contracting. This gaping deficiency is referred to as the *command deficit*, and it is a problem that the punishment argument seeks to correct. In the second and third sections, I defend two premises that together form the punishment argument.

P₁   *Discipline Premise*: Military commanders must be able to discipline soldiers who disobey orders (or who, in some other way, transgress) by imposing severe constraints on the soldiers' freedom.

P₂   *Penal Authority Premise*: The severe constraints on the soldiers' freedom that are sometimes required to preserve military discipline are not constraints that a private agent may justifiably impose. Only public agents should have the moral authority to impose such constraints.

Because private agents may not justifiably impose the constraints on individual freedom that are required for military discipline (P₂) and because military discipline must be assured by commanders (P₁), private agents, I argue, may not exercise command in war. Only public agents may serve as commanders.

While the punishment argument demonstrates that only public agents may exercise command, it does not identify *who* must be placed under the command of public agents. Perhaps, one might argue, a variety of functions may be performed outside the chain of command. The next chapter shows, according to what I call the *control argument*, that *all* risk-imposing agents in war must be placed under a unified chain of command. Together, the punishment and control arguments demonstrate that collective authorizing entities have an obligation to institute a far-reaching publicly controlled chain of command, which would significantly limit the scope of military outsourcing. *If* the demands of these arguments are satisfied, *then*, I argue, rank-and-file military functions may be privatized.

6.1 The command deficit

Let us begin with a shortcoming in modern military discipline that will help frame the punishment argument. The command deficit is a failure by *both* militaries and PMSCs to exercise command responsibility over private contractors. A recent court case, *Saleh vs. Titan Corp.*, illustrates precisely how the US military and PMSCs in Iraq have failed to exercise command responsibility.[6] The case also points to a deep-rooted problem that *any* permissive stance on military outsourcing must confront.

Saleh involved employees of two PMSCs, Titan Corporation and CACI International, who, alongside regular US military personnel in Iraq, allegedly abused prisoners at Abu Ghraib. The two firms were hired to provide interrogation and translation services to US forces at the prison. From October to December of 2003, the firms' employees were among those who committed "numerous incidents of sadistic, blatant, and wanton criminal abuses" according to a US military investigation (*Saleh* Writ 2010: 7; Kinsey 2009: 121).

At the heart of *Saleh* was a question about whether the employees of Titan and CACI were under the command of the companies' corporate managers or whether they were under the command of US military officers. If the employees were under the command of corporate managers, then these managers and ultimately Titan and CACI would be subject to civil liability for the abuses that their employees perpetrated. If, on the other hand, the employees were under the command of US military officers, then Titan and CACI would *not* be subject to civil liability for these abuses. The assignment of civil liability becomes muddied if the US government is found (as it was) to have exercised command over the employees of Titan and CACI. The reason is that the US government claims sovereign immunity in civil court, which means that no party can sue for actions taken by its regular military personnel or by individuals under the command of its regular military personnel. Thus, the assignment of legal responsibility is *asymmetric* in the sense that the US government enjoys immunity for the very exercise of command that would result in the attribution of civil liability to Titan and CACI.

But the assignment of *moral responsibility*, of course, is symmetric. It will track, at least to an extent, our determination of who was in command of the private firms' employees. If the corporate managers of Titan and CACI were in command, then it would be these managers who are subject to the attribution of moral responsibility; if the military officers of the US government were in command, then it would be the officers who are subject to the attribution of moral responsibility. Whether, of course, we actually assign moral responsibility to the entity charged with command over the Titan and CACI employees will depend upon the specifics of how command was exercised.

Underpinning either assignment would be a principle that is familiar to just war theorists—that of command responsibility, which has historically been applied to regular military personnel but may seamlessly be applied to military contractors. According to the principle of command responsibility, a military commander may be morally (and legally) responsible for human rights violations that are committed by soldiers under his charge. He may be responsible for these violations even if the soldiers act without his license.[7] In other words, a military commander must *actively* ensure that his soldiers comply with just war standards. As Walzer (2006: 317) elaborates, he must "see to their training in this regard, issue clear orders, establish inspection procedures, and assure the punishment of individual soldiers and subordinate officers who kill innocent people." Because both military officers and corporate managers may exercise command over private contractors, both entities, in principle, may be charged with command responsibility. That one entity is public and one private should not keep us from insisting that *whichever* entity exercises command on the battlefield (whether public or private) must accept the requirements enshrined in the principle of command responsibility.

But *Saleh* underscores the difficulties that recent efforts at privatization have introduced. Military contractors, like the employees of Titan and CACI, seem to operate within two chains of command, that of the military and that of the corporation. On the one hand, these employees may take orders from military officers, who are ultimately directed by the political leaders and citizens who have authorized war. On the other hand, these employees may take orders from corporate managers, who are ultimately directed by the executives and stock-holders who have contracted to help supply the war.[8] The result is confusion over who is ultimately responsible for the control and discipline of military contractors.

The depth of this confusion underscores a theoretical dilemma that lies at the heart of military outsourcing—and this is the key point that I want to bring out. The US government claimed that military contractors were *not* under their chain of command. A US Army Field Manual insisted: "Management of contractor activities is accomplished through the responsible contracting organization, not the chain of command. Commanders do not have direct control over contractors or their employees … only contractors manage, supervise, and give directions to their employees" (*Saleh* Writ 2010: 77–78).[9]

This attempted delegation of command responsibility would be less troubling if PMSCs took responsibility for the discipline of their employees. However, while firms like Titan and CACI could undoubtedly have taken further measures to effectively supervise their employees (to severely understate the point), these firms possessed limited tools when disciplining contractors and aiming to ensure compliance with just war requirements. Eric Prince, the former CEO of Blackwater, highlighted this very point in his testimony regarding an employee who, while intoxicated, killed an Iraqi civilian: "He didn't have a job with us anymore. We, as a private company cannot detain him. We can fire, we can fine, but we can't do anything else" (*Saleh* Brief 2010: 23).[10] PMSCs lack the apparatus of punishment (for instance, detainment) that states typically reserve for themselves.

The paradox then is that, according to US policy, military officers had "*no* penal authority to compel contractor personnel to perform their duties" while corporations could exercise only a *highly circumscribed* penal authority owing to the limitations that states place (and, I will argue, *should place*) on private punishment (*Saleh* Writ 2010: 78). The consequence is a deficit in which the state neither punishes military contractors nor permits private entities to effectively punish military contractors.

Recent efforts by the US government have attempted to reduce the command deficit. In 2004, Congress expanded the Military Extraterritorial Jurisdiction Act (MEJA) to allow for the prosecution of civilian contractors supporting Department of Defense (DoD) operations (Elsea et al. 2008: 23–24).[11] New congressional legislation (at the time of writing) seeks to extend MEJA to non-DOD operations like those of the State Department (Congressional Record 2014). Alongside these expansions of civilian law, Congress has also placed contractors more fully under the jurisdiction of the Uniformed Code of Military Justice (UCMJ).[12] This enables US military commanders to punish contractors as they would their own soldiers (Elsea et al. 2008: 26; Gates 2008: 2).[13]

But despite these reforms, contractors continue to operate outside of formal command structures (Hedahl 2009: 21; Krahmann 2008: 258; Hammes 2010: 9; Heinecken 2014: 635; Pattison 2014: 146). Amendments to MEJA and the UCMJ strengthen the punishment of errant contractors, but they do not ensure adequate *integration* of contractors into military command. As Charles Tiefer (2009: 755) writes, contractors "are not under military discipline and are not ordinarily acting under the orders of combat commanders." Instead, the US military continues to outsource responsibilities of command to PMSCs.[14]

6.2 Discipline (P$_1$)

The punishment argument objects to this practice by way of two premises: the discipline premise (P$_1$) and the penal authority premise (P$_2$). According to P$_1$, any individual who exercises command must be able to punish soldiers under his command, whether these soldiers disobey orders or misbehave in a way that falls short of explicit disobedience. Moreover, the required punishments may justifiably impose severe limitations on the misbehaving soldiers' freedom.[15] In the first subsection below, I offer a case study of punishment in the US military to guide our discussion. The punishments that US military commanders are permitted to issue—and indeed that military commanders around the world are permitted to issue—are intrusive in the sense that they impose severe limitations on the transgressing soldiers' freedom. But these forms of punishment, I argue, are not too intrusive as to be impermissible. In the second and third subsections, I present two independent rationales for the claim that the penal authority granted to US commanders and to commanders around the world are indeed *needed* to ensure representative and just war. Once these considerations are complete, I ask whether private actors ought to be permitted to dispense (otherwise permissible) intrusive forms of punishment. P$_2$, which is presented in Section 6.3, claims that they should *not* be so permitted.

6.2.1 Non-judicial punishment

US military law, specifically, the UCMJ, enshrines a distinction between three different forms of discipline that I will borrow for my analysis: *judicial punishment*, *non-judicial punishment*, and *non-punitive disciplinary action*. The discipline premise (P_1) focuses specifically on the second of these forms of punishment: non-judicial punishment (henceforth NJP).[16] To understand why P_1 focuses on NJP, we must understand the character of judicial punishment and the character of non-punitive disciplinary action. Judicial punishment is reserved for the most serious offenses and requires legal proceedings in the form of a court-martial. By contrast, both NJP and non-punitive military action enable commanders to discipline their soldiers without formal legal proceedings. What distinguishes NJP from non-punitive disciplinary action is the severity of each form of punishment. NJP includes penalties such as imprisonment and hard labor, while non-punitive military action includes less intrusive measures like counseling and corrective training. As the lawyer David Schlueter (2008: 123) remarks, NJP "serves as a middle ground in the military justice process," providing "sanctions less onerous than a court-martial, yet more severe than non-punitive measures."

The punishment argument takes NJP to be a representative form of intrusive punishment that private entities ought to be barred from imposing. Needless to say, my argument entails that private entities ought to be barred also from imposing judicial punishment (for which the prison sentences are longer and repercussions greater than for NJP). The reason why I leave judicial punishment aside is that private contractors, such as PMSC managers, could in theory refer their employees to the very courts-martial to which regular military personnel are referred. In such instances, private commanders would not be required to adjudicate and sentence personnel within their unit. Instead, they would simply turn over cases to public military judges in just the way that public commanders now turn over cases to public military judges.

The possibility that PMSC managers might turn over cases of NJP to third parties, by contrast (as opposed to judicial punishment), misunderstands the nature of NJP. NJP is premised on the notion that commanders must *themselves* be entitled to punish military personnel in their unit. If this claim is to be plausible, much more needs to be said. I must explain why a third party, such as a military judge or even a commander outside of the unit in question, could not issue punishments to members of that unit and thereby ensure the necessary discipline for the prosecution of just wars. Why, in other words, must rank-and-file soldiers be disciplined by a member of their *own* unit (namely, *their* commander) rather than by someone outside of their unit? This question will be addressed in due course.

For now, I simply wish to stress a definitional point about NJP. Private military personnel serving in positions of command, like PMSC managers, could not refer their employees to NJP by public officers, because NJP is *necessarily* (by definition) a punishment that commanders, public or private, issue to members of their own unit.[17] If a PMSC manager serving as a commander wished to have a member of his unit punished by a third party, the punishment would not qualify as NJP (the punishment would only be NJP if the PMSC manager himself issued it). Given these considerations, NJP will fall within the purview of P_1, while *judicial* punishment will not.[18]

What forms of intrusive punishment are permitted when commanding officers issue NJP? Article 15 of the UCMJ allows for "deprivations of liberty" that include correctional custody, extra duties, restriction, arrest in quarters, and confinement on bread and water.[19] Each form of liberty-deprivation entails a specific set of permissions. Correctional custody may be issued for up to seven days by commanding officers and is defined as physical restraint. Physical restraint requires that the accused be "committed to a designated correctional custody setting" and "placed under the supervision of a monitor" on the understanding that "he or she is not to leave except under specified circumstances" (Schlueter 2008: 152).

Such periods of imprisonment also include "extra duties, hard labor, or other fatigue duties" if so chosen (UCMJ 2015, art. 15). Military courts, unfortunately, have not clearly specified what is meant by "extra duties, hard labor, or fatigue duties." But in practice, these duties include a range of tasks. At the Camp Lejune correctional facility, for example, US Marines serve time on the "rock pile," where they "break big rocks into small rocks," which are then used "for area beautification, such as building and replacing walkways" (Berger 2004: 9). Marines may be compelled to participate in "log drills," which require different manipulations of 14 to 18 foot telephone polls (Berger 2004: 9). Other legally permissible assignments include "repetitively emptying and filling sandbags" and "moving large piles of dirt or heavy rocks for no other purpose than to keep the Soldier working" (Berger 2004: 9).

In addition to correctional custody and such extra duties, NJP may include the liberty-deprivation of "restriction," which means that soldiers are confined to a specific location that is designated by the commander rather than an established correctional facility. When the location of confinement is one's quarters, one is said to be "arrested in quarters." The final type of liberty-deprivation that may be imposed upon soldiers is confinement on bread and water, or diminished rations, for three days. This punishment can only be given to soldiers who are "attached to, or embarked upon, a vessel" (Schlueter 2008: 154). During such confinement on a vessel, soldiers are permitted to communicate only with authorized personnel.

The institution of NJP, dubbed summary punishment in many Anglophone militaries, is not unique to the US (Fidell 2000). In Canada, summary tribunals issue punishments at the level of the unit (National Defence and Canadian Forces 2015; Alleman 2006: 169). The French have a system of NJP but "authorize the local commander considerably more power" than what is given in the US (Gaynor 1953–4: 325). For example, whereas a US unit commander can impose confinement for only seven days, his French counterpart may do so for 15. In the UK, summary punishment by commanding officers may include up to 90 days of detention (UK Army 2015).[20]

Though NJP is ubiquitous in practice, the *need* for NJP on moral grounds is not obvious. What seems clear is that militaries must secure *some* discipline in their ranks and that, if discipline is to be ensured, *some* amount of punishment is needed. If militaries were not permitted to punish disobedient soldiers at all, they would lose a critical deterrent for disobedience. But the hard question for P_1 is not whether some forms of punishment are needed but whether intrusive forms of punishment, like NJP, are needed and whether military commanders (rather than, say, judges) must be permitted to issue these intrusive punishments.

6.2.2 Maximal permissible deterrence

The first rationale that I wish to defend for the proposition that intrusive forms of punishment like NJP must be available for military commanders to discipline members of their unit concerns the need for *maximal permissible deterrence*. Militaries must be able to deter soldiers from disobeying orders for numerous reasons—most notably, as discussed earlier, representation of civilian preferences and respect for *jus in bello*. Moreover, stronger deterrence is preferable to weaker deterrence, and intrusive forms of punishment are better deterrents than non-intrusive forms of punishment. So we have good reason to prefer intrusive forms of deterrence.

But what I am suggesting is something stronger. Given the grave consequences of military disobedience, *particularly* the potential human rights violations that may occur when the moral demands of military discipline are ignored, we ought to insist upon maximal permissible deterrence—i.e., punishments that are maximally intrusive without violating the rights of the punished. Militaries that do not pursue maximal permissible deterrence do not adequately respect the interests of citizens who are being represented, soldiers who serve, and enemy combatants and non-combatants who will otherwise suffer. Of course, some punishments will be *too* intrusive, and so the question arises about whether NJP is problematic on account of its severity. If so, then both public and private entities would be barred from issuing NJP. But if NJP is *not* too intrusive, then militaries have a right and where necessary a responsibility to utilize it as a deterrent.

The debate thus hinges upon whether NJP *is* indeed permissible.[21] One criticism of the practice would challenge the magnitude of sentences that military commanders may issue. But we must acknowledge that such sentences would not be impermissible for *all* crimes. States regularly imprison criminals for more than seven days (the sentence that is permitted by NJP in the US), and prisoners are regularly required to perform onerous tasks while in prison.[22] The criticism would have to be that NJP is unjustifiable (*a*) given the nature of the offense for which such punishments are issued (a proportionality question) or (*b*) given the fact that military commanders are permitted to issue such punishments rather than judges (an authority question). With respect to (*a*), it is certainly true that militaries issue NJP for seemingly minor offenses—inebriation, absence without leave, falling asleep on duty, and so on. But, in a military context, these offenses are *not* minor. Small mistakes in a war zone may hinder a war effort and endanger fellow soldiers and non-combatants. The margin for error is small.

With respect to (*b*), judges may be preferable to military commanders for the dispensation of punishments. And the rigid procedural rules of judges may be preferable to the more flexible rules of military commanders (who, in the vast majority of cases, lack legal training). The problem is that judges cannot accompany military units wherever they go. Moreover, the rigid procedural rules that govern domestic punishment (regarding the collection of evidence, the examination of witnesses, etc.) would likely be too slow and burdensome for discipline in war zones. So, while judges, according to strict procedural rules, should perhaps be charged with issuing *most* of the punishments that ensure military discipline, particularly intrusive punishments whose penalties are akin to NJP, they do not always represent feasible or practicable alternatives to the dispensation of punishments by military commanders. Military commanders are often uniquely positioned to issue these important and otherwise permissible punishments. Though they may not adjudicate as effectively or fairly as judges, the question is whether the value of deterrence that is served by their adjudication outweighs the likely shortcomings. It would seem that they do. If a military commander is unable to threaten NJP and, for reasons of practicality, is incapable of relying upon military judges to issue these punishments, soldiers subordinate to them will have wide latitude for misbehavior. This latitude, I think, would be troubling, to say the least.

A final objection to consider before moving to the second rationale for NJP is that permissible deterrence does not require that the commanding officer of the misbehaving party issue punishment; *some other* commanding officer may issue punishment. If military commanders could achieve maximal permissible deterrence by referring their soldiers for punishment outside of the unit, then private military personnel could serve in positions of command without having to themselves issue NJP. But the bureaucratic obstacles for punishment outside of the unit are likely to impinge upon swift and effective punishment and hence maximal permissible deterrence. In addition, the military commander himself, who is familiar with troops in his command, is likely to have a better grasp than outside commanders over the personalities of the unit and the forms of punishment that are most likely to succeed. Given these considerations, the need for maximal deterrence should prompt us to embrace P_1.

6.2.3 Physically removing soldiers

A second defense of P_1 focuses not upon the need to deter soldiers from violating the rationale for military discipline but upon the need to remove soldiers from battle who are likely to further violate the moral demands of military discipline. In particular, military commanders must be able to imprison soldiers in the way that NJP permits. Soldiers who, through aberrant behavior, come to pose an unjustifiable danger to their military's mission, to their unit, to enemy combatants and non-combatants, or even to themselves, must be removed from battle. And, on occasion, the only way for militaries to assure that this happens when needed is to license military commanders to order the physical detainment of *their* soldiers when they deem fit.

The challenge for this position is to show that (*a*) the power of a military commander to merely fire (rather than detain) transgressing soldiers is insufficient and that (*b*) it must be that military commanders—not just judges or other military commanders—require this power to order imprisonment.

Beginning with (*a*), the reasons we do not want a system in which the military commander possesses the power to fire a misbehaving soldier but not the power to imprison the soldier are numerous. First of all, evacuating a soldier from war rather than detaining him in the war zone may be impossible or imprudent. Evacuations may be financially burdensome and sometimes too dangerous for those who must undertake the evacuation. Second, even when soldiers can be easily evacuated, most militaries cannot afford to lose misbehaving soldiers. To retain a capacity for the prosecution of representative and just wars, militaries may need to keep those soldiers who have committed minor offenses but who can function effectively after being briefly removed from battle. Third, the military relies both upon teamwork and respect for commanding officers. If people were simply evacuated from a war zone for misbehavior, teamwork may be undermined. NJP is a way to preserve unit cohesion while still ensuring that the imprisoned soldier does not pose a danger. Fourth, a military that evacuated misbehaving soldiers from war zones may be providing soldiers with an easy way to get out of combat duties. If all that was required to leave a combat zone was to violate military rules, then militaries may be incapable of effectively retaining and disciplining soldiers. Fifth, war zones are places where soldiers who pose a threat to other members in the unit have, throughout history, been subject to internecine killings—the intentional killing of fellow soldiers. NJP, particularly imprisonment, represents a way to limit this practice. A military commander who lacks the power to punish misbehaving soldiers may not be able to convince those under his command that he will do everything that he can to ensure the preservation of their lives and interests.

But supposing that this is all true, why is it not sufficient for judges or other outside military commanders to detain soldiers? Let us first consider judges and then outside commanders. Regarding judges, as I pointed out above, they cannot accompany all military units into battle. Moreover, even if judges could travel with military units, there is often insufficient time and insufficient resources to have court proceedings in a war zone. Finally, even if court cases for military transgressions were possible, the military expertise of a commander might count as a reason to have the commanding officer be the entity that is permitted to decide whether the soldier must be removed from battle. Military judges may not be good arbiters of whether a soldier poses wrongful danger to his fellow soldiers or to non-combatants. Those who must bear the costs of this danger— the soldiers who suffer when a fellow soldier disobeys orders or misbehaves— may be better attuned to these dangers. If so, the military commander, who is himself risking death and the death of his soldiers, may be preferable to a judge for making decisions about whether a given soldier should be removed from battle or not.

A similar set of concerns applies to military commanders outside of the unit in which punishment is dispensed. The commander of a unit is often detached from other military commanders (just like judges), and he is therefore sometimes the sole entity in position to make determinations regarding punishment of members in his unit. Moreover, the reliance upon outside military commanders to issue punishments like detention will sometimes be burdensome and impracticable (even if possible). Lastly, it will be military commanders themselves who are typically best positioned to determine whether the soldiers under their command ought to be removed from battle, not outside commanders.

To summarize, the need for NJP, and hence the validity of P_1, may be grounded in two ways. The first account relies upon the value of deterrence and the role of intrusive forms of punishment at the level of the military unit to ensure such deterrence. The second account relies upon the value of removing soldiers from the battlefield when they represent dangers to the success of a mission or to other individuals. The power possessed by military commanders to issue NJP serves each of these values and therefore ought to be permitted in war. As the discipline premise (P_1) insists: a military commander must be able to discipline soldiers under his command who disobey orders or who, in some other way, transgress. Such discipline will sometimes require the imposition of severe constraints on the soldiers' freedom, even including imprisonment. These severe constraints, we have seen, are necessary to ensure that wars meet the demands of representation and *jus in bello*. Having now provided a set of considerations to support the first premise of the punishment argument, I will turn to the second premise.

6.3 Penal authority (P_2)

Why must private entities be barred from imposing the intrusive forms of punishment (like NJP) that are necessary for minimal efficacy in the prosecution of just wars? If private military contractors *were* permitted to impose such punishments, then these contractors would be equipped to effectively exercise command (so far as my account is concerned). The punishment argument would thus fail to prohibit private commanders from serving on the battlefield. According to P_2, private military contractors should *not* be permitted to impose intrusive forms of punishment like NJP, because the severe constraints on the soldiers' freedom that are sometimes required to preserve military discipline are not constraints that private agents may justifiably impose (P_2). At least three arguments may be given in defense of P_2. But only the last of these arguments is capable of decisively affirming P_2. If my analysis in this section is correct, then military command must not be outsourced to the private sector.

6.3.1 The conditional argument

The first potentially compelling argument for P_2 is a conditional argument. This argument gains its roots from certain intuitions outside of war. The power to imprison individuals outside of war is reserved exclusively for public institutions (specifically, for judges and juries). Though the management of prisons in states like the US and the UK is increasingly outsourced to the private sector, adjudication and sentencing are not. Robust intuitions seem to underpin the popular insistence that imprisonment—understood as the decision to imprison rather than the actual housing of prisoners—should be reserved for public discharge. If the government must supply *any* services itself, then it seems that the government must furnish a criminal justice system. Put differently, a state that did not alone enforce the law through the adjudication and sentencing of citizens would hardly be recognizable as a state. Our intuitions strongly suggest then that states ought not to hire private companies to adjudicate court cases and sentence those who are found guilty of wrongdoing to prison.

If, according to the conditional argument for P_2, private entities must not be hired to dispense prison sentences outside war, then private entities must not be hired to dispense prison sentences in the military, whether as judges or as military commanders. In both civilian and military contexts, imprisonment severely curtails individual freedom. Insofar as the severe curtailment of freedom is what drives our worry that adjudication and sentencing should not be privately undertaken in a civilian context, it must drive a similar worry in the military context.

That the severe curtailment of freedom *is* what drives our worry about private imprisonment seems plausible. Imprisonment is one of the most substantial liberty deprivations that human beings can impose upon one another. And it seems that there are some liberty deprivations that only states ought to impose (e.g., taxes, conscription, etc.). By contrast, less severe liberty deprivations may be privatized without objection—for instance when a private firm that has been hired by the government to perform road maintenance decides to close down a side street for two hours. According to the conditional defense of P_2 then, the similarity of judges in a civilian context to judges and commanders in a military context (vis-à-vis the liberty deprivation of imprisonment) is sufficient to object to private military commanders.[23] The assumption is that our intuitions regarding the rejection of private civilian judges are correct.

It is because of this assumption, however, that the argument under consideration for P_2 is ultimately unsatisfying. Perhaps, a critic might argue, adjudication and sentencing in the domestic context *need not* be publicly undertaken. Why precisely should we care if private entities were permitted to impose substantial liberty deprivations? If the government were to outsource criminal justice and hire private firms to decide court cases—private firms that specialized in adjudication and sentencing—perhaps, the critic might continue, fairer decisions would be reached. The first argument for P_2 thus appeals to strong intuitions but does not explain *why* criminal adjudication ought to be a public enterprise.

6.3.2 *The liberty forfeiture argument*

Having attempted to ground P_2 with an argument that is conditional upon robust but undefended intuitions, let us turn to an argument that aims to defend these intuitions. The liberty forfeiture argument claims that there are some liberties that *only* public actors may justifiably curtail, and the liberty deprivations of NJP, particularly imprisonment, are examples of such liberty curtailments. To see the force of this argument, we ought to distinguish at the outset between two types of liberties that are often taken to be inviolable: liberties that *nobody* may justifiably curtail and liberties that *nobody but public actors* (e.g., states) may justifiably curtail.

Those who insist upon the existence of the first type of liberty often point to slavery as a paradigmatic example. Their contention is that individuals are not permitted to sell themselves into slavery, because to do so would entail a violation of *inalienable* rights. For proponents of this view, there are certain acts X that are wrongful if committed against an individual *even* when that individual has consented to X. Others, by contrast, will be tempted to insist that, so long as individuals consent to such practices under fair bargaining conditions, then they *are* justified in submitting to severe liberty deprivations. The resolution of such questions is difficult. Ultimately, however, the liberty deprivations that they probe are less germane to our analysis than the second set of liberty deprivations—after all, we typically think that *some* actors, namely states, are permitted to imprison people.

Deprivations of the second kind of liberty take the following form: there are certain acts Y that are wrongful if committed against an individual by a private entity, even when that private entity has consented, but are *not* wrongful when committed by a public entity. Here, imprisonment is the paradigmatic example. It is thought to be justifiable if publicly imposed but unjustifiable if privately imposed (such a divergence, of course, is what P_2 aims to demonstrate).

If P_2 is to be defensible according to the liberty forfeiture argument, it must be shown that consent to future NJP is an example of Y: individuals are justified in joining the military and thereby consenting to future imprisonment by public actors but not in joining, say, a PMSC and consenting to future imprisonment by PMSC employees. The question is: why? Assuming that all of the usual requirements of consent are satisfied—that it is freely given, according to fair procedures, with sufficient information, and so on—why would it be wrong for PMSC employees and PMSC managers to sign a contract that permits the PMSC to imprison its employees? Why not simply insist that governments must regulate the punishments that are dispensed by private commanders and thereby protect the rights of those who are punished?

Perhaps the most promising approach to distinguish between the individual who consents to public punishment and the individual who consents to private punishment focuses upon the right to democratic participation that can be exercised by the prisoners of a government. When an individual consents to future imprisonment by private contractors, that individual is, in a sense, *wholly* giving up her liberty. Once imprisoned by a private contractor, this individual would lack a guaranteed say in the terms of her imprisonment. By contrast, if this person were to consent to future imprisonment by public agents, she would not be *wholly* giving up her liberty. So long as the public actor was a democratic government, she would still retain a vote in collective decisions about how citizens (including her) are punished. Thus, if I were to join the US military and be imprisoned under the UCMJ, I would still be able to vote in US elections and thereby voice a say over the terms of my imprisonment (and the terms of all other imprisonments).

But an opponent might maintain that private contractors simply do not give up their liberty in the way that I have described. If, for instance, I owned stock in a publicly traded PMSC and was commanded by an employee of that PMSC, then I would retain a say in how the PMSC punished me as a stockholder (whether I was an independent contractor, an employee of the PMSC in question, an employee of a different PMSC, or a publicly employed soldier). So long as I owned stock, then I would have a right to contribute at shareholder meetings about how the PMSC should punish.

However, I would lack a *guaranteed* say in the decisions of the PMSC. An important difference between being a stockholder and a voting citizen is that one always retains a democratic vote as a citizen.[24] One is precluded from voluntarily giving up the meta-liberty of alienating input over the liberty deprivations that may permissibly be imposed upon oneself and other citizens. By contrast, stockholders can always sell their shares. Moreover, stockholders may be bought out without their consent (for instance, if the publicly traded company were transformed into a privately traded company). Thus, as a PMSC stockholder, since I might unavoidably lose my shares, there are no guarantees that I will retain a say over the terms of my imprisonment. Perhaps then, it is the guaranteed say that distinguishes private imprisonment from public imprisonment.

But three responses decide against this position—and therefore against the liberty forfeiture argument for P_2. First, a government might simply legislate against the undemocratic nature of private punishment. For instance, the government might require that all private contractors serve in PMSCs that are democratically run (e.g., worker cooperatives) and be punished only by other members of the PMSC. In this case, one would retain a say over the terms of one's imprisonment.

Second, even if we assume that a right to vote can only be truly *guaranteed* in a public institution, it is still not clear whether retaining the right to vote in prison really safeguards one's liberty. Does a prisoner who casts a vote in elections retain *meaningful* control over the terms of her imprisonment in a way that individuals who are imprisoned by private actors, operating in the context of a well-regulated legal system, lack?

Finally, even assuming such meaningful control, the libertarian will rightfully press us to explain why individuals should not be free to consent to future forms of punishment in which they forfeit a meaningful say over the terms of their punishment. Perhaps the notion that an individual would be permitted to sell himself into permanent slavery would (and should) continue to make us uneasy. But the sorts of liberty-deprivations under consideration—the imprisonment and hard forms of labor of NJP—are *temporary*. Any serious offense, which might entail years in prison or even a lifetime, would be referred to a court-martial or civilian court and would not be adjudicated by military commanders. As we saw above, the prison sentences of NJP are on the order of days, not months or years. Thus, the libertarian critique of the liberty forfeiture argument for P_2 remains unanswered. Why should individuals not be free to consent to private NJP-like punishments by military commanders? And why should governments who wish to rely upon private commanders not be permitted to honor such contracts and permit private commanders to punish?

6.3.3 The conflict-of-interest argument

A third, and I believe decisive, argument for P_2 focuses upon conflicts-of-interest as a source for discrediting private judges and commanders. The principle that underpins the third argument is encapsulated in the legal doctrine of *nemo judex in sua causa*. No man, as Alexander Hamilton (1996 [1788]) summarized, ought "to be a judge in his own cause, or in any cause in respect to which he has the least interest or bias."[25]

One of the most important applications of this principle in contemporary legal systems is to financial interests. A judge who possesses a legal stake in a business that comes before her court is typically disqualified from hearing the case. For instance, if I own even a single share of stock in Facebook, then, as a judge in most jurisdictions, I would be prohibited from deciding cases in which Facebook was involved. But financial interests do not just include legal ownership (Shaman 1989–90: 281). A judge who lacks stock in Facebook but sits on its board or who is employed by Facebook may possess financial interests in the success of the company that preclude her from hearing a case that involves Facebook. The worry, of course, is that a judge who stands to benefit financially from a given decision will be more likely to make that decision than a judge who does not stand to benefit. The mechanism by which bias infiltrates is complex—whether the judge consciously promotes her own financial interests or subconsciously interprets evidence in such a way that her financial interests are promoted. Regardless, the tendency for this to occur should be sufficient for a society to insist that its judges *not* have financial interests in the cases that they must adjudicate.

The problem of outsourced judges in the civilian context extends readily to commanders in a military context. Private contractors who are permitted to exercise military command would seemingly possess conflicts-of-interest when punishing soldiers under their command. In particular, they would possess incentives to punish in ways that increased their likelihood of winning future contracts—in contrast to regular military commanders who do not have to vie for future contracts. Because these conflicts-of-interest undermine military discipline and justice, private commanders, I argue, must not be employed.

To understand the potential conflicts-of-interest that may arise for private contactors in war, we should remember that all military contractors are either independent contractors or employees of a PMSC.[26] The former refer to those who contract directly with the entity that has authorized war, while the latter refer to those who serve as employees of a private firm, which, in turn, contracts with the entity that has authorized war. In serving as military commanders, independent contractors might exercise command over public military personnel, other independent contractors, or PMSC employees. Similarly, PMSC employees might exercise command over public military personnel, independent contractors, or PMSC employees (whether in their own firm or other firms). With each command relationship, different conflicts-of-interest may apply.

Such conflicts-of-interest may be usefully broken down into two categories. The first concerns the command relationship between PMSC employees and other members of their own PMSC. The second concerns all other command relationships. Regarding the first class, PMSCs like Blackwater have strong incentives for *their* employees to be spared punishment. A PMSC whose employees were regularly being punished for disobedience or violations of military code is unlikely to win future contracts. Thus, a Blackwater manager who commands other Blackwater employees in war is likely to be unduly lenient, even when severity better serves military discipline and justice. Through lenience, the manager would preserve the reputation of his PMSC.

This criticism, however, only applies to PMSC employees who are commanding individuals from *their* PMSC. It does not apply to PMSC employees who command employees of other PMSCs, regular military personnel, or independent contractors. Nor does it apply to independent contractors who exercise command, whether over PMSC employees, regular military personnel, or other independent contractors. In these cases, the military commander would possess no disincentive to punish soldiers in his unit on account of joint-allegiance to the same PMSC.

But other conflicts-of-interest would exist. On the one hand, it may be that private contractors who command individuals from outside of their PMSC—a necessary condition to keep contractors from being lenient on employees in furtherance of their PMSC's reputation—will be unduly severe in punishing subordinates. After all, subordinates may be competitors in the future, whether in winning future contracts (if the soldiers are PMSC employees or independent contractors) or gaining future work (if the soldiers are regular military personnel). Supposing that severity in punishment would indicate errant behavior in the same way that leniency in the case above would indicate good behavior, then private commanders may use harsh forms of punishment to diminish the reputations of subordinates. To the extent that military contracting functions as a zero-sum game, in which case the diminished reputation of one private contractor would translate into the enhanced reputation of the next, the problems of competition may be particularly corrosive to military discipline and justice.

On the other hand, it may be that these incentives will point in the opposite direction, compelling private contractors to collude with one another rather than compete. Military contractors may believe that excessive punishment will reflect poorly on *all* private contractors and may thus exercise undue leniency.[27] Just as the indictment of a politician for fraud is often seen as an indictment of all politicians, the misbehavior of some private contractors may be viewed as an indictment of the private security industry as a whole. For that reason, private contractors that command other private contractors may be hesitant to punish one another—and, by the same token, *eager* to punish regular military personnel. These incentives of competition and collusion may point in one direction in some circumstances and in the opposite direction in other circumstances. In either case, the incentives will undermine military discipline and justice.

At least three objections to this argument must be addressed. The first two will be treated together. First, the incentives that have just been identified of collusion and competition may be thought to *balance out* with one another. Since the commander possesses both incentives and disincentives to punish, any potential conflicts-of-interest seem to dissolve. The commander would have no reason to punish one way or another. A second way to criticize the conflict-of-interest argument is to maintain that, regardless of the balance between incentives and disincentives for punishment, militaries could, in a sophisticated way, manipulate these incentives. To prevent leniency, militaries could reward punishment with money, benefits, or future contracts. To prevent stringency, military could do the opposite.

The problem is that both of these criticisms miss the corrosive nature of conflicts-of-interest. We do not want military commanders to be *thinking about* the financial incentives or disincentives of punishment. Commanders should punish fairly, as they see fit, to facilitate the prosecution of representative and just wars. To assure this, financial interests should be divorced from the dispensation of punishment. This is why, in the civilian context, a judge who owns stock in Facebook would be disqualified from a case even if certain incentives pointed in a direction that would *disadvantage* Facebook—for instance, if the opponent of Facebook in the case were the judge's best friend. Judgment may be clouded despite the fact that financial incentives cancel out, so to speak. It is also why, in a civilian context, governments do not offer incentives to judges who exhibit fairness in cases where a conflict-of-interest exists. Instead, governments simply insist that judges recuse themselves.

A third criticism of my argument is that regular military personnel *also* have financial interests in doling out appropriate levels of punishment. They are seeking to rise up through the ranks of the military (much like private contractors are trying to win future contracts). Indeed, in both civilian and military courts, judges must decide cases knowing that their decisions may hinder or harm their careers as judges. Those judges who punish in one way may be promoted to a higher court, while those that punish in a second way may not be promoted. Judges therefore have a strong incentive to punish in a way that would enable them to be promoted to a higher court.

To see how this third objection misses the mark, however, we should consider an important distinction between public agents and private agents. Public agents, as we saw in Section 5.3.2, are bound by a set of maximizing institutional strictures to *ensure* that they promote the collective interests of those whom they serve, while private agents are bound by something less. The difference between public judges and private judges, in other words, is that public judges are constrained by the full menu of safeguards to *ensure* fair adjudication in furtherance of collective interests: public judges are typically paid salaries; they are not personally liable for actions taken in their judicial auspices; they receive pensions and health care; they are eligible for awards and honors that are reserved for governmental employees; they wear special uniforms. All such efforts help alleviate conflicts-of-interest. Private judges are not bound by these maximizing institutional strictures.

Thus, I do not deny that public judges face certain conflicts-of-interest—after all, judges are also people in the world with careers to pursue (though I would contend that these conflicts are far less acute than those faced by private judges). But public judges are shielded from these conflicts by a set of maximizing constraints that ensure sound decision-making. And justice, I believe, requires that every effort be made to neutralize the potential for conflicts-of-interest. Otherwise, the severe liberty constraints that result from such decisions look to become illegitimate and the rights of the accused violated.

Based upon this rationale, military commanders must not be private agents. To ensure fair adjudication by commanders, military authorizing entities must apply the full menu of safeguards to their commanders, just as they would apply them to judges. The rights of the accused demand fairness and objectivity in judgment that would go unprotected if commercial interests were permitted to shape the punishment-practices of commanders. Moreover, the rights of immune risk-bearers on or near the enemy side may be undermined: lax or poorly functioning punishment mechanisms, which would exist to the extent that conflicts-of-interest clouded the judgment of commanders, could jeopardize effective command responsibility and hence the achievement of representative and just war. To adequately protect the rights of military personnel who are accused of misbehavior and to protect immune risk-bearers who must often bear the horrific costs of malfunctioning military command, governments must make every effort to remove conflicts-of-interest from their commanders and thereby facilitate appropriate punishment.

6.4 Conclusion

In light of the punishment argument, military command must not be outsourced to the private sector. P_1 demonstrated that a military commander must be able to discipline soldiers who disobey orders (or who, in some other way, transgress)—sometimes using severe forms of punishment embodied in NJP. P_2 then showed that, to avoid unjustifiable conflicts of interest, punishments like NJP must not be issued by private agents. When the punishment argument is tethered to the governance argument, the scope of impermissible outsourcing is widened: the responsibilities of commanders *and* their superiors must be publicly discharged.

Notes

1 See also Section 1.7 for my definition of command. Note that my definition does not include a prohibition on private command. This chapter asks whether private actors *should* be permitted to exercise command (over regular soldiers and/or other contractors). To assume that, by definition, they cannot exercise command would be to foreclose further analysis.

2 The largest unit over which individuals exercise command is the company, which is typically an organizational unit of between 48 and 250 soldiers (or three and five platoons). But while the unit over which individuals exercise command may be as large as a company, it may also be as small as a squad (just eight or nine soldiers). It will depend upon whether the unit is attached or detached to a larger organizational unit. For instance, if a squad is attached to the company, then the company leader will be the commander, while the squad leader will be his subordinate. If the platoon is detached from the company, then the leader of the platoon will be the commander. As the United States Army (2011: 5) summarizes, "whether an officer is a commander is determined by the duties that he or she performs, not necessarily the title of the position occupied."

3 I use the terms "commander" and "commanding officer" interchangeably.

4 It is possible that, in a sufficiently small military, the functions of high-ranking military officers and commanders may be carried out by the same set of individuals. This is not a problem for the governance or punishment arguments, which insist only that *some* set of public agents fulfill these functions.

5 As Schwenkenbecher (2013: 161) highlights, others argue for a criterion of legitimate authority that requires authorizing entities to be capable of enforcing obedience to *jus in bello*. See Thompson (2005), Held (2005), and Finlay (2010).

6 The US Court of Appeals for the District of Columbia Circuit heard this case. For the most detailed account, see *Saleh et al. vs. Titan et al.*, Writ of Certiorari (2010). References to this writ will henceforth be referred to as *Saleh* Writ (2010).

7 This point is not to suggest that US military commanders were unaware of the abuses that were being perpetrated at Abu Ghraib or that US military commanders did not participate in these abuses. The opposite was true. See Hersh (2004: 1–73).

8 See Hedahl (2009: 24–25) for worries about bringing contractors under the military's chain of command *and* worries about leaving them outside.

9 Note that the PMSCs claimed that the employees *were* under the military's chain of command. But, according to US policy, they were not.

10 Walzer (2008: 21) points out the troubling reality that even those contractors who are fired may be rehired by other PMSCs.

11 MEJA, initially passed in 2000, allowed for the prosecution of contractors directly hired by the DoD, not all contractors supporting DoD operations.

12 Prior to 2007, contractors could only be tried under the UCMJ in times of "declared war." Senator Lindsay Graham proposed an amendment to a defense spending bill passed in 2007 that extended the UCMJ's reach to contractors during "contingency operations" as well as declared war. See Witte (2007).

13 See Shah (2014: 2559–2560) for helpful criticism of the UCMJ amendment.

14 The delegation, it must be noted, is informal. Formally, "a civilian, other than the President as Commander-in-Chief (or National Command Authority), may not exercise command" (United States Army 2014: 1).

15 "Intrusive forms of punishment" are understood simply as forms of punishment that impose severe limitations on the transgressing soldiers' freedom. The nature of such punishments will be spelled out below.

16 NJP is known as "Article 15" in the Air Force, "Office Hours" in the Marine Corps, and "Captain's Mast" in the Navy and Coast Guard. It is known as summary punishment in militaries outside of the US. See Schlueter (2008: 123).

17 Under US military law, NJP is reserved for public military officers. But, of course, since this chapter probes whether private commanders ought to be permitted to issue NJP (or NJP-like punishments), the possibility is left open that both public and private commanders might be entitled to issue NJP. Private military personnel, in other words, are not, *by definition*, prohibited from issuing NJP.

18 The reason why non-punitive disciplinary action is excluded from my analysis is because the mildness of its penalties mean that their dispensation by private entities is much less objectionable than NJP (if it is objectionable at all). We might say that non-punitive disciplinary measures are scarcely different from the penalties that a football coach might impose upon his players for showing up late to practice—for instance, extra running. Since the aim of this chapter is to identify a set of harsh punishments that militaries enable their commanders to impose but that private entities must not be permitted to impose, NJP will serve as the focal point of analysis, and non-punitive disciplinary action and judicial punishment will be left aside.

19 There are four categories of NJP: reduction in grade, forfeiture of pay, censure, and deprivation of liberty (Schlueter 2008: 152). Since the first three forms of punishment are akin to punishments that are routinely dispensed in the private sector, and unobjectionably so, only the final form of punishment is explored here.

20 Because NJP plays such a prominent role in the punishment argument and because NJP is a distinctively military practice, a question arises about whether the punishment argument could apply to contractors besides those hired by the DoD: for example, those hired by the Central Intelligence Agency (CIA), the State Department, or the Justice Department. To the extent that contractors are supporting DoD operations (to use the language of MEJA), then the punishment argument may apply. However, whether it applies more broadly will depend upon whether the agencies in question possess any mechanisms of punishment comparable to NJP under the UCMJ.

21 It may be that forms of punishment that are more intrusive than non-judicial punishment are permissible and that they should be preferred to NJP to achieve maximal permissible deterrence. But this only strengthens my argument. To the extent that harsh forms of punishment (e.g., long prison sentences) are permissible, we have a class of punishments that may be particularly problematic for private entities to impose—though permissible for public entities to impose.

22 One might think that the unpleasantness of tasks that are performed in prison pales in comparison to the unpleasantness of tasks that are performed in the military. Still, few would maintain that the "hard labor" required in the military—filling up sandbags, digging ditches, and so on—would be impermissible to impose upon, say, mass murderers. It is just that these punishments may seem inappropriate in the context of domestic punishment, where discipline is not an end under pursuit like in the military.

23 A related objection would focus on military judges. *If* military judges must be public agents, which seems plausible, then surely military commanders must be public agents since both must issue punishments within the same criminal justice system.

24 I leave aside the question of whether individuals who commit certain crimes should ever forfeit their right to vote. Such crimes are typically of a magnitude that exceed the crimes for which NJP is imposed and thus fall outside the scope of my analysis.

25 John Locke (2002 [1690]: ch. 2) is perhaps the first to make this argument:

> it is unreasonable for men to be judges in their own cases, that self love will make men partial to themselves and their friends: and on the other side, that ill nature, passion and revenge will carry them too far in punishing others.

See Harel (2011: 400) for more on this.

26 See Section 5.1 for more on these definitions.
27 Caparini (2008: 174) points to a telling story of two former employees from the PMSC Triple Canopy who sued the company, "claiming they were fired for filing a report on their supervisor whom they accused of shooting at Iraqi vehicles and civilians without provocation, and that they were subsequently 'blacklisted' and prevented from finding work with other PMSCs operating in Iraq."

References

Alleman, Lindsy Nicole (2006). "Who is in Charge, and Who Should Be? The Disciplinary Role of the Commander in Military Justice Systems," *Duke Journal of Comparative and International Law*, Vol. 16, No. 1, pp. 169–191.

Berger III, Joseph B. (2004). "Making Little Rocks Out of Big Rocks: Implementing Sentences to Hard Labor Without Confinement," *The Army Lawyer*, Department of the Army Pamphlet 27–50–379, pp. 1–21. Available at www.loc.gov/rr/frd/Military_Law/pdf/12-2004.pdf (accessed December 28, 2015).

Caparini, Marina (2008). "Regulating Private Military and Security Companies: The US Approach," in Andrew Alexandra, Deane-Peter Baker, and Marina Caparini (eds), *Private Military and Security Companies: Ethics, Policies, and Civil–Military Relations* (New York: Routledge), pp. 171–188.

Congressional Record (2014). "Statements on Introduced Bills and Joint Resolutions." Vol. 160, No. 109. Available at http://fas.org/irp/congress/2014_cr/s2598.html (accessed December 28, 2015).

Elsea, Jennifer K., Moshe Shwartz, and Kennon Nakamura (2008). *Congressional Research Service Report for Congress: Private Security Contractors in Iraq: Background, Legal Status, and Other Issues.* Available at https://fas.org/sgp/crs/natsec/RL32419.pdf (accessed December 28, 2015).

Fidell, Eugene R. (2000). "A World-Wide Perspective on Change in Military Justice," *Air Force Law Review*, Vol. 48, pp. 195–209.

Finlay, Christopher J. (2010). "Legitimacy and Non-State Political Violence," *The Journal of Political Philosophy*, Vol. 18, No. 3, pp. 287–312.

Gates, Robert (2008). *Memorandum for Secretaries of the Military Departments, Chairman of the Joint Chiefs of Staff, Under Secretaries of Defense, Commanders of the Combatant Commands.* Available at www.fas.org/sgp/othergov/dod/gates-ucmj.pdf (accessed December 28, 2015).

Gaynor, James (1953–4). "The French Code of Military Justice: A Comparison with the Uniform Code of Military Justice," *George Washington Law Review*, Vol. 22, pp. 318–336.

Hamilton, Alexander (1996 [1788]). "The Federalist, No. 10," in William Brock (ed), *The Federalist* (London: J.M. Dent).

Hammes, TX. (2010). "Private Contractors in Conflict Zones: The Good, the Bad, and the Strategic Impact," *National Defense University Strategic Forum*, No. 260. Available at http://psm.du.edu/media/documents/reports_and_stats/think_tanks/inss_hammes-private-contractors.pdf (accessed December 28, 2015).

Harel, Alon (2011). "Outsourcing Violence," *Law & Ethics of Human Rights*, Vol. 5, No. 2, pp. 395–413.

Hedahl, Marcus (2009). "Blood and Blackwaters: A Call to Arms for the Profession of Arms," *Journal of Military Ethics*, Vol. 8, No. 1, pp. 19–33.

Heinecken, Lindy (2014). "Outsourcing Public Security: The Unforeseen Consequences for the Military Profession," *Armed Forces & Society*, Vol. 40, No. 4, pp. 625–646.

Held, Virginia (2005). "Legitimate Authority in Non-State Groups Using Violence," *Journal of Social Philosophy*, Vol. 36, No. 2, pp. 175–193.

Hersh, Seymour (2004). *Chain of Command: The Road from 9/11 to Abu Ghraib* (New York: HarperCollins Publishers).

Kinsey, Christopher (2009). *Private Contractors and the Reconstruction of Iraq* (Oxford: Routledge).

Krahmann, Elke (2008). "The New Model Soldier and Civil–Military Relations," in Andrew Alexandra, Deane-Peter Baker, and Marina Caparini (eds), *Private Military and Security Companies: Ethics, Policies, and Civil–Military Relations* (New York: Routledge), pp. 247–265.

Locke, John (2002 [1690]). *The Second Treatise of Civil Government* (Austin: Liberty Library). Available at www.constitution.org/jl/2ndtreat.htm (accessed December 28, 2015).

National Defence and Canadian Forces (2015). "Military Justice at the Summary Trial Level." Available at www.forces.gc.ca/en/about-reports-pubs-military-law-summary-trial-level/index.page (accessed December 28, 2015).

Pattison, James (2014). *The Morality of Private War* (Oxford: Oxford University Press).

Saleh et al. vs. Titan et al., Writ of Certiorari (2010). Available at http://ccrjustice.org/files/Saleh%20v%20Titan%20cert%20petition.pdf (accessed December 28, 2015).

Saleh et al. vs. Titan et al., Brief of *Amici Curiae* Retired Military Officers in Support of Petitioners (2010). Available at http://ccrjustice.org/files/05.28.10%20Brief%20of%20Retired%20Military.pdf (accessed December 28, 2015).

Schlueter, David (2008). *Military Criminal Justice: Practice and Procedures*, seventh edition (Albany: LexisNexis Matthew Bender).

Schwenkenbecher, Anne (2013). "Rethinking Legitimate Authority," in Fritz Allhoff, Nicholas G. Evans, and Adam Henschke (eds), *Routledge Handbook of Ethics and War: Just War Theory in the Twenty-First Century* (Oxford: Routledge), pp. 161–170.

Shah, Reema (2014). "Beating Blackwater: Using Domestic Legislation to Enforce the International Code of Conduct for Private Military Companies," *The Yale Law Journal*, Vol. 123, No. 7, pp. 2559–2573.

Shaman, Jeffrey M. (1989–90). "Bias on the Bench: Judicial Conflict of Interest," *Georgetown Journal of Ethics*, Vol. 3, pp. 245-290.

Tiefer, Charles (2009). "No More Nisour Squares: Legal Control of Private Security Companies in Iraq and After," *Oregon Law Review*, Vol. 88, No. 3, pp. 745–776.

Thompson, Janna (2005). "Terrorism, Morality, and Right Authority," in Georg Meggle (ed), *Ethics of Terrorism & Counter-Terrorism* (Frankfurt: Ontos Verlag), pp. 151–160.

Uniform Code of Military Justice (2015). Available at www.au.af.mil/au/awc/awcgate/ucmj.htm (accessed December 28, 2015).

United Kingdom Army (2015). "Summary Hearings." Available at http://spa.independent.gov.uk/test/about_us/military_criminal_ju.htm (accessed December 28, 2015).

United States Army (2011). *Military Justice: Army Regulation 27–10* (Washington, DC: Headquarters, Department of the Army). Available at www.apd.army.mil/pdffiles/r27_10.pdf (accessed December 28, 2015).

United States Army (2014). *Army Command Policy: Army Regulation 600–20* (Washington, DC: Headquarters, Department of the Army). Available at www.apd.army.mil/pdffiles/r600_20.pdf (accessed December 28, 2015).

Walzer, Michael (2006). *Just and Unjust Wars: A Moral Argument with Historical Illustrations*, 4th edition (New York: Basic Books).

Walzer, Michael (2008). "Mercenary Impulse: Is there an Ethic that Justifies Blackwater?," *The New Republic*, March 12, pp. 20–21.

Witte, Griff (2007). "New Law Could Subject Civilians to Military Trial," *The Washington Post*, January 15.

7   Control

The previous two chapters have advanced a theory of public military supply. The responsibilities of the *military's leadership*—consisting of commanding officers and their superiors on the chain of command—must be publicly discharged. But we have yet to examine the roles that are occupied by the vast majority of men and women who participate in war: the *rank-and-file* personnel. The rank-and-file consist of those individuals who serve outside the military leadership. They include lower-level military officers who lack command responsibilities (when, for instance, these officers are attached to a larger organizational unit), enlisted personnel (since none exercise command), and contractors who discharge the responsibilities of enlisted personnel.[1]

Typically, the rank-and-file vastly outnumber the military leadership. As an approximation of this disparity, consider the ratio of enlisted military personnel to officers, which is 4.7 to 1 in the US military (United States Office of the Deputy Assistant Secretary of Defense 2013: 11) and 5.7 to 1 in the UK military (United Kingdom Ministry of Defence 2014: 5). These numbers are coarse approximations since lower-level officers who do not exercise command are counted as members of the rank-and-file on my account, as are military contractors who discharge the responsibilities of enlisted personnel. Thus, the ratio of rank-and-file personnel to military leaders will be even greater than these numbers suggest—which should only underscore the point that the previous two chapters addressed a very circumscribed set of military functions and neglected difficult questions about the vast majority of military personnel.

In this chapter, I shift attention to this more expansive group. Rank-and-file military personnel certainly lack the governing responsibilities of high-ranking officers and the disciplinary responsibilities of commanders. Yet, many are charged with tasks that are commonly taken to be ineligible for justifiable outsourcing. In particular, rank-and-file military personnel do most of the killing in war.[2] If the function of killing is one that ought not to be outsourced, then the account of public military supply that I have put forward in the previous two chapters is incomplete. More generally, if any further moral commitments, besides those that were wielded in Chapters 5 and 6 to challenge military outsourcing, are inconsistent with privatization in the rank-and-file, then my account will be wanting.

This chapter argues that the responsibilities discharged by rank-and-file military personnel may be privatized *only if* contractors are placed under a uniform chain of command.[3] I call this the *control argument* for institutional integration. The control argument demands more than enhanced regulation. It applies even if an alternative regulatory scheme could deliver the same outcomes on the battlefield.

My argument builds on the work of Cécile Fabre (2012: 214–218). We should recall from Chapter 5 that Fabre's position on private military supply is a conjunction of two claims: the weapons-claim and the army-claim. According to the weapons-claim, "if a political community is at liberty to buy guns from private manufacturers" in order to carry out just defensive killings, then "it is also at liberty to buy soldiering services from those willing to provide them" (Fabre 2012: 215). According to the army-claim, "if a political community has a right to pay for a standing army—as it surely does—then it also has a right to pay for a private army" (Fabre 2012: 216). Up to this point, I have challenged the army-claim by identifying a set of functions that ought to be discharged publicly. But I have left the weapons-claim intact. The question to now take up is: might the weapons-claim at least permit the privatization of rank-and-file functions, even if it cannot permit the privatization of leadership functions?

I will argue that, if properly amended, the weapons-claim *can* serve as a justification for privatization in the rank-and-file—though the amended claim sets substantial limits on privatization, which Fabre would reject. The first section sketches Fabre's weapons-claim. I explain how a regulatory framework defended by Deane-Peter Baker (2011a: 122–143) helps to flesh out her conception of proper institutional regulation (Fabre 2012: 227–228, 238). The second section considers a challenge to the weapons-claim that applies even in the presence of sound institutional regulation: James Pattison's (2014: 90–97, 103–106, 164–165) "problem of private choice." Though this challenge rightfully questions the permissibility of delegating discretion to contractors on the battlefield, it does not account for how a chain of command might adequately limit this discretion. The third section turns to a defense of formal command structures offered by Avihay Dorfman and Alon Harel (2013).[4] Their defense reaches the right conclusion about integrating contractors into a chain of command but does so according to an unjustifiable rationale. The fourth section then presents the control argument in place of Dorfman and Harel's unsuccessful defense. I show that contractors must be integrated into the chain of command to avoid violating the risk-prevention argument (of Part I). The control argument thus delivers the following revision of the weapons-claim, which is defended in the fifth section: *if* authorizing entities place contractors under a uniform chain of command, then their reliance upon private actors to kill (and support killing) in war may be justified according to the same rationale that the purchase of weapons is justified. The upshot of these considerations is that a collective authorizing entity may privatize its rank-and-file so long as it meets the conditions of the control argument.

Before moving on to these arguments, we should note that the requirement defended in this chapter would, if instituted, substantially alter the current landscape of military outsourcing. As we saw in Section 6.1, despite recent efforts towards enhanced regulatory oversight, governments continue to employ contractors outside of formal command structures (Hedahl 2009: 21; Krahmann 2008: 258; Hammes 2010: 9; Heinecken 2014: 635; Pattison 2014: 146). One of the great appeals of PMSCs, from the pragmatic vantage of governments who rely upon them, is that they free regular militaries to devote manpower to core combat competencies (the "tooth-to-tail ratio") while entrusting contractors to *independently* complete tasks (Pattison 2014: 15). The requirement that militaries place battlefield contractors under a uniform chain of command would likely eliminate many of the perceived strategic and financial benefits of outsourcing.

But the depth of integration envisaged by this chapter must be distinguished from a more demanding alternative. According to the more demanding alternative, contractors would form military units like the French Foreign Legion or the UK Brigade of Gurkhas. These units are comprised of foreign nationals who are recruited, trained, armed, paid, and deployed by the French and UK militaries (respectively) in a manner scarcely distinguishable from regular soldiers. Members of these units, unlike private contractors, are bound to serve only one military and fight wars authorized only by that military's government.

My proposed arrangement demands something less. It has more affinity with the UK Sponsored Reserves (SR). The SR is a program in which the government contracts with private companies that provide employees to serve as reservists in the UK military. The UK Ministry of Defense explains that "[c]ontinuing civilian employment and pay whilst undertaking SR service is based on the concept of an employee being on loan to a third party" (United Kingdom Defence Forum 2015). Private contractors on my model, like those in the SR, would serve within the military's formal command structures, receiving orders either directly from regular military personnel or from contractors serving under the command of regular military personnel. But unlike the typical contracts that govern SR service, contracts need not be signed ahead of military authorization; they may be signed once war has been initiated (and they may be limited so that contractors are free to serve a different military in the next war). I simply insist that contractors be incorporated into a chain of command as they are in the SR.

One final point to record before moving forward is that my argument does not show that governments *ought to* outsource any military functions.[5] First of all, as my discussion thus far has made clear, current institutional structures—which regulate training, command structures, accountability, and so on—require improvement. Second, even with dramatically improved control of the sort envisaged by this book, other considerations speak against military outsourcing. In particular, governments may have strategic, economic, and legal reasons to forego military outsourcing.[6] Lastly, military outsourcing may not cohere with the values of some societies.[7] I am simply defending the *right* of governments to outsource rank-and-file responsibilities, not its advisability.

7.1 The weapons-claim

Let us turn then to the weapons-claim of Fabre's (2012: 214–218) argument for military privatization. Fabre, we may recall from Section 5.2, deploys a four-step argument to defend the weapons-claim, beginning from the plausible presupposition that individuals have a positive right to receive *some* assistance from third parties in order to lead minimally flourishing lives (Fabre 2007: 363):

a   Based on such a positive right, one who is starving to death must be entitled to receive food from a third party if the costs to the third party are sufficiently low (Fabre 2007: 366).
b   But if one is entitled to receive food from a third party in order to avoid starvation, then one must also be entitled to receive a gun from a third party to defend oneself (Fabre 2007: 367).
c   If one is entitled to receive a gun from a third party to defend oneself, then one must also be entitled to receive assistance from a third party to thwart attack. The rationale that would permit the former would also permit the latter—namely that material resources and services are both "fungible and scarce" and sometimes necessary "to pursue our ends" (Fabre 2007: 366).
d   But if one is entitled to receive assistance from a third party to thwart attack, then one should also be entitled to offer incentives like money to garner this assistance (for instance, in the way that patients pay surgeons for operations) (Fabre 2012: 220).

As we can see from (*a*) to (*d*), the weapons-claim maintains that we are justified in receiving, and indeed purchasing, killing services in the same way that we are justified in receiving and purchasing food and guns. If this justification for private killing could *further* motivate a justification for the private performance of other, non-combat rank-and-file military functions, then outsourcing in the rank-and-file would be justifiable.

The difficulty is that the weapons-claim, as stated, does not seem to proceed unobjectionably from (*b*) to (*c*). In one crucial respect, the purchase of guns is *very* different from the purchase of killing services. When an individual purchases a gun, she retains control over how the gun will be used and hence how the killing will proceed. By contrast, when she purchases a killing service, she loses (at least some) control over how the killing will proceed. Of course, if this loss of control is to undermine the weapons-claim, it must be a *morally problematic* loss of control. Otherwise, the permission to purchase a gun could still ground the permission to purchase a killing service. But it should prompt us to reflect upon the moral implications of discretion that may be sacrificed when one hires a service-provider to use material tools (e.g., guns) on one's behalf rather than using the material tools oneself.

Fabre (2012: 227–228, 238) aims to mitigate this worry by stipulating that the weapons-claim assumes sound regulation of private actors. Though she does not elaborate the features of proposed regulation, she cites others who do.[8] Among those she cites, Deane-Peter Baker (2011a: 85–143; 2011b) offers perhaps the most comprehensive regulatory framework—one rooted in Peter Feaver's (1996; 1998; 1999; 2003) work on agency theory in civil–military relations.[9] According to agency theory, relationships between buyers (principals) and sellers (agents) are best understood as strategic games of players with competing motivations. The two central problems of agency theory are adverse selection and moral hazard: the former describes the predicament of principals being unable to "know for certain about the true preferences and capabilities" of an agent before hiring her; the latter describes the difficulty of principles being unable to know for "certain whether the agent is working or shirking" once hired (Feaver 2003: 73–74). Baker maintains that a tightly regulated system of military outsourcing can overcome both of these problems and thereby neutralize concerns about contractor discretion.

Though he does not explicitly extend this framework to Fabre's move from (*b*) to (*c*), the logic of such an extension is straightforward. For Baker (and Fabre), the problems of adverse selection and moral hazard, need not be more threatening to (*c*) than (*b*).[10] Adverse selection may not be threatening for two reasons. First, it applies no less forcefully to the purchase of guns in (*b*) as it does to the purchase of killing services in (*c*). One cannot know, before purchasing a weapon, whether the weapons-manufacturer is, for instance, honest and meticulous, just as one cannot know, before purchasing a killing service, whether the killer is skilled and patient. In both cases, the principal must, *on some level*, trust that the agent will act as expected. Second, whether in purchasing guns or hiring killers, the principal can easily mitigate the problems of adverse selection. The principal might run background checks on the agent, get references, complete interviews, conduct personality tests, inspect prior work records, and, ideally, rely upon a licensing system that performs assessments centrally (and regularly) so that all licensees meet a threshold of reliability. In these ways, a buyer may be able to purchase weapons or killing services without fear that adverse selection will necessarily invalidate the killings that these purchases facilitate. It may even be, assuming proper safeguards, that the prospect of an individual purchasing a weapon and carrying out a killing himself is *more* worrying than the prospect of this individual simply hiring a professional—the individual may be a very incompetent killer. In other words, the problem of adverse selection may be *less alarming* than the problem of an incompetent principal (with regard to his capacity for killing).

The problem of moral hazard, on first glance, may appear to be more troubling for the weapons-claim. Principals have methods of neutralization at their disposal in (*b*) that simply cannot be wielded in (*c*). It is true that the principal in (*b*) will encounter *some* moral hazard since he cannot know whether the gun-seller will produce the weapon according to specification. This could jeopardize the success of just killings; it could potentially endanger bystanders, if the gun's accuracy was sufficiently distorted; and it could compromise the proportionality of the killing, if the gun's firepower was over-amplified. But the principal has an easy way to counteract such moral hazard: test the gun before use. That way, the errors (or wrongs) of the gun-seller are detected *ahead* of any killings, and the

gun-seller is rendered somewhat (though not entirely) impotent to act in opposition to the moral commitments of the principal vis-à-vis the killing. By contrast,
for the principal in (c), no such tests are available. Whereas the "behavior" of a
gun can largely be assured ahead of its use, the behavior of a hired killer, even
one chosen scrupulously to avoid adverse selection, cannot. Hired killers possess
autonomy, which guns lack, and may thus act in opposition to the moral commitments of the principal in a way that guns cannot.

Still, without further argumentation, this concern is unlikely to be decisive in
distinguishing the killing in (c) from the killing in (b) such that only the latter is
justifiable. The reason is that, like adverse selection, moral hazard may be coun-
teracted with thick regulation. Principals may draft detailed and clearly specified
contracts that outline the obligations of the agent. The principal may monitor the
agent, whether himself or via a third-party. Breaches of contract may result in
financial penalties, loss of eligibility for future contracts, and other forms of punishment.[11] In these ways, the principal in (c) may be able to ensure that the agent
acts in accord with his (the principal's) moral commitments and desires with
respect to the killing.[12]

If we are to challenge Fabre's move from (b) to (c) in the weapons-claim, we
must explain why, following application of thick regulation, the residual discretion that remains in (c) and not (b) is morally problematic. Moreover, we must
explain why only private actors, and not regular military personnel, possess this
problematic residual discretion.

7.2 The problem of private choice

James Pattison (2014: 90–97, 103–106, 164–165) does both in defending the
problem of private choice. His defense, if successful, explains why the discretion
in (c) is morally problematic and why it is uniquely problematic for military contractors. The objection takes aim at a key difference between private actors and
regular military personnel when carrying out killings. No matter how effectively
private actors are regulated—even if a military institutes every safeguard that
Baker identifies—*still* private actors may, in the end, choose not to provide their
services. They may refuse to participate in war *before* signing a contract (by
turning it down) and *after* signing the contract (by violating its terms).[13] In
neither case can the military compel performance as it can with regular military
personnel. Contractors thus possess a level of potentially objectionable discretion that regular military personnel lack. And this may explain why the justification for purchasing weapons in (b) cannot ground the employment of contractors
in (c) (even if it can ground the employment of regular military personnel).

Pattison offers three reasons ("deeper objections") to explain why the discretion of private actors is objectionable.[14] First, a state's *ability to prosecute just
wars* may be undermined (Pattison 2014: 90–92). The government may not be
able to authorize just wars if candidate contractors decline risky commitments
(*ad bellum*), and it may not be able to execute just wars if contractors decline
risky missions (*in bello*). Second, reliance upon contractors may frustrate the

democratic dictates of a community in its decision to initiate war (since contractors may not sign up) and in decisions related to the conduct of war (since contractors may not cooperate) (Pattison 2014: 103–106). Finally, the problem of private choice could result in the *under-provision* of military protection (Pattison 2014: 164–165). If contractors refuse to participate in certain wars, then they could leave whole states vulnerable to security threats; if they refuse to participate in certain missions, then they could leave specific populations vulnerable. On all three grounds—efficacy, democracy, and distributive justice—the problem of private choice explains why the purchase of private (though not public) killing services in war is problematic in a way that the purchase of guns is not. Governments that rely upon contractors, even with robust regulation, cannot *ensure* obedient participation. By contrast, they can ensure the obedient participation of regular military personnel.[15]

But Pattison's argument may be challenged in two ways. The first concerns his insistence that private actors cannot be compelled to sign contracts (*agreement*), and the second concerns his insistence that private actors cannot be compelled to keep contracts (*compliance*). Together, these weaknesses undermine his defense of the problem of private choice as an objection to military privatization.

7.2.1 *Agreement to contracts*

Beginning first with agreement to contracts, Pattison does not properly acknowledge how governments *can and do* compel private entities to participate in war just as they compel citizens to participate in war. Governments often nationalize private companies in emergency circumstances when these companies cannot or will not provide goods and services required by the military.[16] Why should the services of PMSCs or private contractors be any different?

Pattison's argument is premised on the plausible notion that, if some good *g* is sufficiently valuable, governments ought to reserve the right to *ensure* that *g* is provided. For instance, if farmers were unwilling to grow food for soldiers or factory workers were unwilling to manufacture jets, the government may be required to compel these farmers and factory workers to provide their services. But just because the government must reserve the right to nationalize private companies in times of war does not mean that it must do so *ahead* of war. The fact that private farmers or factory workers may refuse to sign contracts with the military to provide food or jets, in other words, does not alone suffice as an argument for nationalizing the agriculture and aeronautics industries.

It may not even undermine the cumulative legitimacy of these industries.[17] The disadvantages of waiting until emergencies arise before instituting nationalization—on measures of efficacy, democracy, and distributive justice—may be outweighed by the advantages of waiting (while keeping these industries private) *based on the very measures* that Pattison endorses (of efficacy, democracy, and distributive justice). A government that nationalizes its agriculture and aeronautics industries only in times of emergency may do as well (or better) in providing effective and democratically sanctioned military force for all citizens than a government that nationalizes these industries ahead of war. It will all depend upon whether the government is adept at farming and manufacturing (or at least of managing farmers and manufacturers).

The same line of reasoning applies to PMSCs and private contractors. The government may nationalize PMSCs just as it does other companies in times of emergency if they are unwilling to accept contracts. Likewise, the government may compel individuals to sign contracts to provide services for the military just as it compels them to join the regular forces.[18]

Pattison might insist that the problem of private choice at least diminishes the cumulative legitimacy of private contractors since governments could more rapidly (and easily) initiate wars with regular military personnel. But, as with farming and manufacturing, this will all depend upon how capable the government is of providing the services on offer by the private sector. A government may find that its military power is *enhanced* through deferral of certain tasks to innovative PMSCs. It may be able to more effectively, more democratically, and more equitably fight wars *all-things-considered* by hiring PMSCs when these companies will sign contracts while reserving the right to nationalize them in time of emergency.

Even if this were true, however, one might argue that the problem of private choice at least applies to governments that rely upon *foreign* PMSCs and contractors rather than *domestic* PMSCS and contractors. Suppose that the Canadian government were to depend entirely upon Blackwater for the prosecution of its wars. The Canadian government could not nationalize Blackwater in times of emergency, because Blackwater is not a Canadian company. But this counterexample only underscores that the Canadian government ought to make certain that *someone* (whether public or private) be available for compulsion in times of emergency, not that public actors must alone be available. All that we can insist upon, based on the problem of private choice, is that a government ought *either* to field its own military *or* develop a private sector of military services to rely upon in times of emergency. We cannot conclude that the government ought to choose the former over the latter without further argumentation that a system of private contracting and emergency nationalization diminishes military efficacy, democracy, and distributive justice more than a regular military force.[19]

7.2.2 Compliance with contracts

The second way in which the problem of private choice fails concerns *compliance* with contracts (rather than agreement to contracts). Pattison argues that contractors cannot be compelled like regular soldiers to carry out their duties. But the question is: why not? Why not simply place contractors under the military's chain of command? If contractors consented to being placed under the chain of command when signing agreements with collective authorizing entities, surely collective authorizing entities would be justified in placing them under the chain of command.

Pattison (2014: 90–91 fn 20) is resistant to this retort, arguing that it becomes "questionable whether there could exist such a system without the differences between private contractors and regular soldiers being somewhat eroded" (the difference "is usually held to relate to the latter being under the authority structures of the regular military and the former not"). But the inclusion of contractors in the chain of command is certainly consistent with Pattison's (2014: 17) definition of private contractors (and mine) and with internationally established definitions (e.g., the Montreux document [International Committee of the Red Cross 2009: 9]).[20] Recall the operative definition of private contractors in this book: following Fabre (2012: 210), they are individuals who supply military services to "belligerent[s], against payment, outside the state's military recruitment and training procedures, either directly to a party in a conflict, or through an employment contract" with a PMSC. The employees of a PMSC who are recruited, trained, equipped, and paid by a PMSC to serve under the formal command structures of a belligerent would continue to meet this definition. Employees of this PMSC would still be recognizable as private contractors and would continue to evoke many of the criticisms that they now evoke.[21]

The normative case for integrating private contractors into formal command structures then cannot be vulnerable to the definitional argument that integration would erode the distinction between contractors and regular military personnel. If morality requires that contractors be included in the chain of command, then the distinction ought to be eroded. The question is whether contractors *indeed ought to be included*. More must be said to ground such a requirement of inclusion for it to be plausible. But we cannot conclude based upon the problem of private choice that contractors possess a level of discretion that regular military personnel lack. In the end, governments may compel contractors to perform their duties—through nationalization in times of crisis and the chain of command during war.

Thus, Fabre's analogy between the purchase of weapons in (*b*) and killing services in (*c*) remains in tact. Without further argumentation, the weapons-claim cannot be rejected according to a purported asymmetry of discretion between contractors and regular military personnel. The discretion of *both* contractors and regular military personnel may be restricted. If the purchase of weapons by a government is morally analogous to the employment of regular military personnel, then it may also be morally analogous to the employment of contractors.

7.3 Fidelity of deference

Having considered one objection to the weapons-claim, let us now turn to Dorfman and Harel's (2013) case for integrating contractors into the chain of command. Like Pattison, they recognize that the discretion of contractors on the battlefield may be problematic. However, unlike Pattison, they insist that contractors *should* indeed be incorporated into formal command structures.[22] Also unlike Pattison, they take the benefit of formal command structures to concern the *guidance* rather than *compulsion* of rank-and-file personnel. For Dorfman and Harel, the problem with employing contractors outside of formal command structures is not that militaries would be unable to compel them to perform their duties. It is that contractors would lack the necessary guidance to make sound decisions in battle.

Dorfman and Harel defend this position by way of a hypothetical scenario. Imagine that Rex, a benevolent dictator, is responsible for arresting criminals, adjudicating court cases, and administering the death penalty. Rex may delegate his administration of the death penalty in one of three ways. The first is by grabbing an assistant's hand "and pressing the fingers of this person against the gun's trigger" as though the fingers "are the natural extensions of Rex's human body" (Dorfman and Harel 2013: 72, 73). If Rex were justified in receiving a gun from a third party to administer the death penalty, then, on this conception of assistance, it seems that he must also be justified in receiving killing services from a third party (where Rex retains complete control over whether and how the trigger is pulled). This is an extreme and, of course, fanciful example of how the purchase of killing services in (*c*) might be morally indistinguishable from the purchase of weapons in (*b*).

The two other forms of delegation grant further discretion to the assistant and are distinguished from one another by the mode of fidelity that each requires: whether *fidelity of deference* or *fidelity of reason*. An assistant who administers the death penalty with fidelity of deference suppresses her own judgment and acts entirely according to Rex's judgment. By contrast, an assistant who exercises fidelity of reason does not suppress her own judgment but rather aims to execute Rex's decision fairly and in accord with the general good. Whereas the mark of successful delegation on the former conception of fidelity is whether the assistant acts based upon *Rex's reason*, the mark of successful delegation on the latter conception of fidelity is whether the assistant acts based upon *right reason*. Only fidelity of reason entails value judgments about the task that is performed.

According to Dorfman and Harel, killing in war requires that soldiers act by fidelity of deference. This is because when soldiers kill by fidelity of reason, they "subject their potential victims to their judgments of how to proceed with the war" and thereby "assume a normative power that individual persons normally lack, namely, the standing to subject other human beings to their private judgments, including judgments concerning the justness of killing and maiming them" (Dorfman and Harel 2013: 98). Soldiers must suppress their own judgments so that every act on the battlefield may be attributed to the collective judgment of the state, not the personal judgment of individual soldiers. Only in this way can individuals avoid violating the dignity of others on the battlefield.

The linchpin of this argument is that certain institutional structures are needed for individuals to *know* what fidelity of deference of requires. Dorfman and Harel propose two criteria. The first criterion is that an agent must act within a community of practice: "an ongoing framework or coordinative effort in which participants immerse themselves *together* in formulating, articulating, and shaping a shared perspective from which they can approach, systematically, the implementation and execution of government decisions" (Dorfman and Harel 2013: 82). The second criterion is that political leaders must provide ongoing input to the community of practice to help shape its decisions. Governments that rely upon agents to kill in war outside of this two-pronged framework unjustifiably compel rank-and-file personnel to subject enemies to their own personal judgments about killing in war.[23] Thus, contractors must be incorporated into integrative communities of practice—like formal command structures—to limit their discretion. Doing so renders their killings more like the killing performed by Rex's assistant when Rex pulls the assistant's fingers against the gun's trigger (and indeed more like the purchase of a gun if Rex himself were to deploy it).

The problem with this argument is that fidelity of deference should *not* be the guiding principle of rank-and-file personnel on the battlefield.[24] Soldiers must take responsibility for their actions and decide for themselves whether the orders that they receive are just or unjust. Though rank-and-file personnel may be excused for obeying unjust orders and disobeying just orders given the profound coercion that they experience on the battlefield, they should nevertheless be expected to *attempt* these judgments rather than suppress them. Dorfman and Harel (2013: 100–101) acknowledge that their position flies in the face of revisionist just war theory, which sets a high bar for personal decision-making; they also acknowledge that it may challenge traditionalist just war theory, which sets a lower bar but nevertheless requires disobedience to patently unjust orders.[25] The demand that soldiers suspend their judgment on the battlefield largely removes the capacity for soldiers to satisfy even this lower bar of disobedience. And this removal, I think, is unjustifiable.[26]

7.4 The control argument

We have now seen that the chain of command may serve as a mechanism of compulsion (Section 7.2) and as a conduit of information about collective preferences (Section 7.3). The control argument emphasizes *both* functions in defending the chain of command and accounts for the individual right (and obligation) to resist compulsion and ignore collective preferences when orders are unjust. My defense of the chain of command—which accommodates the political obligations of rank-and-file personnel to promote collective preferences and the moral obligations of rank-and-file personnel to promote human rights—paves the way for a revision of the weapons-claim in this section.

The control argument begins from the presupposition (defended at length in Sections 4.1–4.3) that rank-and-file personnel who participate in war impose *some* risks (direct and indirect) upon the fundamental interests of individuals who have done nothing to forfeit their freedom from substantial risk-imposition. Moreover, if risk-imposition is to be justifiable, risk-bearers must have a say in decisions of military authorization and continuation (according to the risk-prevention and governance arguments). The control argument insists upon one further requirement: collective authorizing entities must place risk-imposing agents under a uniform chain of command to ensure that agents avoid imposing risks that have *not* been collectively authorized. The argument runs as follows:

P₁   *Unauthorized risk-imposition*: If a member of the rank-and-file imposes risks that have *not* been collectively authorized by relevant risk-bearers, then that individual imposes risks upon individuals who lack a say in her decision of risk-imposition, which is unjustifiable.

P₂   *Transmitted justification*: Only a uniform chain of command allows members of the rank-and-file to impose collectively authorized risks.

If P₁ and P₂ are sound, then a collective authorizing entity must place risk-imposing agents under a uniform chain of command. Otherwise, the collective wrongfully relies upon individuals who lack justification to impose risks that accompany killing and maiming in war. By the same token, individuals who agree to kill and maim outside of a formal command structure behave wrong-fully since they impose risks without proper justification.

7.4.1   Unauthorized risk-imposition

Let us consider each premise in turn. I argued in Chapters 3 and 4 that individuals who *authorize* military force without consulting risk-bearers violate their fundamental interests in physical security. The claim of P₁, by contrast, is that rank-and-file personnel who contribute to military *supply* in the absence of consultation with risk-bearers violate their fundamental interests in physical security. Of course, as I acknowledged in Section 5.5.2, the risks that an individual soldier imposes *in bello* are far less weighty than the risks that an authorizing entity imposes when unleashing war or that a high-ranking military officer imposes when directing a war. Individual *in bello* risks are unlikely to attain the magnitude needed to trigger the risk-prevention or governance arguments and therefore need not be imposed by public officials alone on account of those arguments. But P₁ maintains only that rank-and-file personnel impose *some* risks upon the fundamental interest of others.

They impose these risks *directly* by participating in the infliction of collateral damage during attacks.[27] This is true not only of those who provide killing services but also those who contribute to acts of killing—for example, individuals who maintain weapons, drive vehicles, stand guard, and so on. These agents bear *some* causal responsibility for the harms that are imposed upon immune risk-bearers, including combatants and civilians on or near the enemy side and fellow rank-and-file members subjected to "friendly fire." Indeed, the mere presence and movement of rank-and-file personnel in the physical space of war creates opportunities that they will come under attack, which, in turn, may prompt them to participate in defensive action that imposes direct risks upon the fundamental interests of others who possess immunity.

Moreover, rank-and-file personnel impose *indirect risks* upon one another and their civilian affiliates in war. Rank-and-file personnel may reveal the positions of fellow combatants through careless behavior, thus prompting enemy attack. They may take a wrong turn, requiring the launch of a rescue operation that imposes risks on fellow rank-and-file fighters. Mistakes may even alter the course of war so that new risk-imposing offensives are undertaken. Consider the four Blackwater employees who were escorting a kitchen supply convoy in 2004 and were kidnapped and hung over a bridge in Fallujah (the contractors were allegedly not provided with maps or logistical support) (Taussig-Rubo 2009: 125–126). This prompted the infamous First Battle of Fallujah (the American Operation Vigilant Resolve), which resulted in numerous American and Iraqi casualties. We can never know, of course, what would have happened had the Blackwater employees *not* been kidnapped. But this case illustrates that rank-and-file personnel who operate on the battlefield can make mistakes (or commit wrongs) that prompt future risk-imposing retaliatory behavior. When authorizing entities fight wars on their home soil, this retaliatory behavior may have profound consequences for civilian affiliates on or near the battlefield.

These acts of direct and indirect risk-imposition, like all acts of risk-imposition, require justification. That war has been collectively authorized does not mean that rank-and-file personnel who participate in supply enjoy a green light to impose risks as they deem fit. Rank-and-file personnel are also moral agents who must uphold the rights of others. Yet, unlike other moral agents, their justification for risk-imposing behavior is not likely to be acquired via personal consultation with risk-bearers. The battlefield is not the kind of place where rank-and-file personnel can consult with innocent bystanders across enemy lines, for example, before firing bullets or launching air strikes. They must instead rely upon *previously made* decisions by the collective that has authorized (and is continuing) war. If rank-and-file personnel are to justifiably bypass battlefield consultation with risk-bearers, they must impose risks according to these collective decisions, because only the collective has consulted with risk-bearers and determined the appropriate distribution of risks.

7.4.2 *Transmitted justification*

What conditions must be satisfied, if any, for rank-and-file personnel to bypass battlefield consultation? The answer, according to P_2, is a uniform chain of command. Though rank-and-file personnel may be unable to personally consult with risk-bearers in a war zone, they *can* (and must) communicate with their superiors on the chain of command, and these superiors can (and must) ultimately communicate with the collective authorizing entity. Because collective authorizing entities reflect the preferences of risk-bearers (according to the risk-prevention and governance arguments), they provide rank-and-file personnel with a justification for their risk-imposing behavior by transmitting orders down the chain of command.[28]

An immediate objection to consider, however, is that the chain of command is not needed for this transmission. Why would contracts not suffice? If a collective authorizing entity hired private actors to impose risks according to the terms of an agreement, then that agreement would provide contractors with the needed justification for their risk-imposing behavior so long as the behavior complied with agreed-upon guidelines. The vehicle for transmitting collective preferences to risk-imposing agents, on this model, would simply be the contract itself rather than the chain of command.

But the problem with this objection is that decisions of risk-imposition are ongoing in war and require continuous interaction between the authorizing entities and their risk-imposing agents. Contracts are underspecified and cannot easily anticipate all collective decisions that will be needed.[29] Once hostilities begin, collectives face unforeseen questions about *how* and *upon whom* to impose risks. Moreover, as wars spread, new risk-bearers acquire entitlements of participation in collective decision-making. Static contracts cannot reflect how the preferences of risk-bearers shift nor how the preferences of new risk-bearers may alter collective decisions. Therefore, absent a chain of command, even contractors who strictly abide by the terms of their agreements may nevertheless impose risks that have not been chosen by the collective, whether these risks deviate from collective preferences or simply fall outside the scope of what the collective has considered.

Robust oversight, while helpful, would not be sufficient. The collective not only requires knowledge about how its soldiers and contractors impose risk but also the ability to *direct* contractors in real-time to ensure that risk-imposition is consonant with collective preferences. Such a two-way system of ongoing monitoring coupled with authoritative orders *is* a chain of command and is precisely what the control argument requires. It demands a conduit, in continuous operation, between collective decisions that respond to the vagaries of war and individual acts of risk-imposition in the field.

7.4.3 Two objections

With the control argument complete, we should consider two objections before moving to the revised weapons-claim. The first objection is that the control argument is of limited applicability since contractors do not *always* impose risks in war, particularly if they do not participate in combat. Perhaps contractors could serve outside the chain of command without falling foul of the control argument. This objection, in one sense, is correct. As the contributions of contractors to risk-imposition shrink, the need for their incorporation into a chain of command weakens. This may explain why private actors who grow the food that soldiers eat or sew the uniforms that they wear ought to be excluded from formal command structures. Indeed, one of the strengths of the control argument is that it provides framework, based on individual contributions to risk-imposition, for assessing who may be excluded from command structures and who must be included.[30] Insofar as contractors participate in direct and/or indirect risk-imposition of war, they *must* be included in the chain of command.

A second objection is that rank-and-file personnel, whether contractors or regular soldiers, must not be expected to impose *unjust* risks, even if these risks reflect collective preferences. This is precisely the objection that I raised against Dorfman and Harel in Section 7.3. Soldiers must make judgments about the justice of their actions and not *merely* defer to collective dictates when imposing risks.

But notice how the requirement that soldiers disobey unjust orders is not threatening to the control argument. I have defended the chain of command as a mechanism for collectives to provide rank-and-file military personnel with justifications for their risk-imposing behavior on the battlefield. Without collective justification, rank-and-file personnel impose risks upon the fundamental interests of others despite providing risk-bearers with no say in their decisions. But, if the orders that rank-and-file personnel receive on the battlefield are unjust, then rank-and-file personnel cannot rely upon these orders to justify their risk-imposition. Instead, the regular soldier or contractor must act like any individual thrust into a situation that requires self-defense or other-defense. That is, she must make difficult personal choices about risk-imposition. Though she will lack access to collective justifications through the chain of command, she may consult with risk-bearers on the battlefield if the situation permits. Otherwise, there are times when she must accept the grave implications of disobedience, including her own death, in order to avoid imposing unjustifiable risks on others.[31] Rank-and-file personnel have both political responsibilities to impose risks that have been collectively determined *and* individual moral responsibilities to refrain from imposing risks that they deem unjustifiable.

7.5 The revised weapons-claim

Having challenged two important objections to the control argument, we may now return to the weapons claim with the conceptual tools at our disposal to defend an appropriate revision. The revised weapons-claim is this: *if* authorizing entities place contractors under a uniform chain of command, then their reliance upon private actors to kill (and support killing) in war may be justified according to the same rationale that the purchase of weapons is justified.

The rationale that may justify the purchase of weapons *and* the purchase of killing services derives seamlessly from the compensation argument for military authorization (defended in Section 4.5). Collectives have an obligation to protect those who have been disarmed through public monopolization. In order to provide military protection, collectives must have the capacity to purchase (or produce) weapons and employ individuals to prosecute wars. Material resources and services, as Fabre (2007: 366) insists, are both "fungible and scarce" and sometimes necessary "to pursue our ends." If a public monopolizing entity is to properly compensate those who have been disarmed, as it must, the monopolizing entity must procure weapons *and* military services for the prosecution of collectively authorized wars.

As we have seen, the primary difference between the purchase of weapons and the purchase of killing services is that individuals who sell the latter retain far more discretion over killings than individuals who sell the former. But the control argument limits this discretion by placing risk-imposing agents, both public and private, under a uniform chain of command. This ensures, *contra* Pattison, that contractors can be compelled to carry out tasks that reflect collective preferences and that they receive appropriate guidance regarding what collective preferences demand. *Contra* Dorfman and Harel, the control argument does not obliterate discretion by demanding fidelity of deference. It requires that rank-and-file personnel rely upon the chain of command to justify their risk-imposing behavior *if* (and *only if*) the orders are just. Otherwise, in the case of unjust orders, rank-and-file personnel ought to disobey. Indeed, disobedience is precisely what we should expect of weapons manufacturers who are asked to produce guns for unjust wars. In both the sale of weapons *and* military services, individuals have duties to promote justice.

7.6 Conclusion

In this chapter, I have defended a particular conception of formal command structures. These structures are needed to provide rank-and-file personnel with a justification for their risk-imposing behavior on the battlefield. Once a chain of command is instituted, the weapons-claim (on a revised understanding) becomes more plausible. The purchase of weapons *is* like the purchase of military services. What underpins the moral justification for the purchase of weapons and military services is the duty that collective monopolizing entities possess to protect those who have been disarmed (according to the compensation argument).

This chapter thus finalizes a set of restrictions that ought to limit the privatization of military supply. A collective authorizing entity must employ public agents to carry out the functions of high-ranking military officers (according to the governance argument) and commanders (according to the punishment argument). Moreover, it must place all risk-imposing agents under a uniform chain of command (according to the control argument). *If* these requirements are satisfied, then a collective authorizing entity may privatize the functions of its rank-and-file personnel.

In light of the analysis in this chapter, we may reject A_5—which entails exclusively public authorization and exclusively public supply. Because A_0–A_2 were rejected in Chapters 3 and 4 and A_3 was rejected in Chapters 5 and 6, our rejection of A_5 in this chapter leaves only A_4 for endorsement: military force must be authorized publicly but *may* be supplied privately or publicly so long as it satisfies the requirements of the governance, punishment, and control arguments.

Notes

1 In conscripted armies (without enlistment), the rank-and-file would include conscripts outside of the military leadership and contractors who discharge the responsibilities of conscripts (if any were hired).
2 Outsourcing killing responsibilities runs counter to US policy, for instance, which demands that roles not be privatized if they "significantly affect the life, liberty, [and] property" of other persons (Daniels 2003).

3 I use the terms "chain of command" and "formal command structures" synonymously. A "uniform chain of command" is simply a *single* chain of command with the collective authorizing entity at the top.

4 Dorfman and Harel (2013) offer the clearest expression of a position defended in Harel (2008, 2011, 2014: 65–106). I cite Dorfman and Harel (2013) except when material is present only in Harel's other work.

5 For similar caveats in the philosophical literature on privatization, see Fabre (2012: 210–211).

6 See Section 8.1 for an overview of many current problems with private military contracting.

7 As a US citizen, I believe that military outsourcing runs counter to some of the values that the US has long embraced. This is why I believe that we should elect political leaders who are opposed to extensive military outsourcing, even though governments may have a right to engage in the practice.

8 Fabre (2014: 228 fn 25) cites Baker (2011a: 122–143), Avant (2005: 143–177), O'Brien (2007: 29–48), Krahmann (2007: 94–112), and Singer (2007: 230–242).

9 See also Singer (2007: 151–168) for a discussion of privatization framed in the language of agency theory. Singer is less optimistic than Baker that contractors may be sufficiently regulated.

10 Baker, it should be noted, seems committed to Fabre's move from (*b*) to (*c*) given his broad endorsement of her argument (Baker 2011a: 80, 161).

11 Note that contracts, monitoring, and punishment are, of course, also available to the gun-buyer in (*b*) when making a purchase from a gun-seller.

12 Indeed, as we saw above with adverse selection—where a professional killer might be better qualified than the principal himself when it comes to killing—it may be that a professional killer is *better equipped* to follow the moral commitments of the principal than the principal himself. Weakness of will is no small matter in killing. As a result, we should recognize that even the principal in (*b*), who purchases a weapon to perform the killing, may act in opposition to his own considered preferences and moral commitments.

13 As Taussig-Rubbo (2009: 136) reminds us, "the contractor can quit," but "the solder who leaves can, under certain circumstances, be executed for desertion." See also Singer (2007: 159–162), Wulf (2008: 195), Wolfendale (2008: 219), Kasher (2008: 240), and Krahmann (2008: 248) for similar worries.

14 Each is a "deeper objection" to military privatization that hinges on the problem of private choice. Recall from Section 5.2 that deeper objections, according to Pattison, differ from contingent objections in that they apply *even in the presence of* perfect regulatory efficacy. Though Pattison does not believe that deeper objections always decide against the moral permissibility of outsourcing, he believes that they detract from its cumulative legitimacy. Along with the three deeper objections to privatization considered here, Pattison proposes two further deeper objections to privatization (based upon contractor motivations and communal bonds), which I take up in Chapter 8.

15 One might be tempted to criticize the problem of private choice by insisting that it underestimates the capacity of incentives to guide human behavior. Pattison seems to presume that there are some exceptionally risky contracts for which no amount of money will suffice to entice individuals to serve. While this presumption is, I believe, correct, it is important to note that the problem of private choice does not depend upon it being correct. Governments cannot always afford the high prices of military contracts. Reliance upon contractors might therefore leave governments vulnerable even if contractors were willing to accept high payouts for brave, exceptional risks.

16 In 1918, for instance, when the workers of a large American gun manufacturer, Smith & Wesson, threatened to strike, potentially delaying the production of weapons, the US government nationalized the company (Jinks and Krein 2006: 75).

17 For Pattison, the legitimacy of military action is cumulative in the following sense: if military action fails to achieve one parameter that contributes to overall legitimacy, the authorizing entity "may be able to make up for this loss and still possess an adequate degree of legitimacy, all-things-considered, by doing well on other qualities" (Pattison 2014: 73). The qualities under consideration here are efficacy, democracy, and distributive justice.

18 Even governments with all-volunteer forces (like the US) typically reserve the *legal right* to conscript their citizens. In the US, all men aged 18–25 must register through the Selective Service System, which is tasked with instituting a draft in times of emergency.

19 If the government is incapable of establishing either a public military or a private sector of military services, then it may not satisfy the principles of military efficacy, democracy, and distributive justice. But this violation will owe to the government's lack of resources, not the problem of private choice.

20 Pattison (2014: 17), we should recall, defines PMSCs as "private firms that provide military and/or security services that involve or assist the use of force beyond the borders of their own or their client's political community."

21 These criticisms will be explored in Chapter 8.

22 Note that Dorfman and Harel (2013) do not refer specifically to formal command structures but rather, as we will see, to "integrative communities of practice." I take the chain of command to represent the most plausible integrative community of practice that might apply in the military.

23 It is important to note that, according to Dorfman and Harel (2013: 86), contractors cease to be "private agents" once they are incorporated into integrative communities of practice like the military's chain of command. They become public agents. Just as Pattison (2014: 90–91 fn 20) argues that contractors who operate within formal command structures are not exactly *contractors* (as we understand them), Dorfman and Harel argue that they are not exactly *private agents* (though they still count as contractors). But, unlike Pattison, Dorfman and Harel insist upon integrating contractors into these formal command structures.

24 See Pattison (2014: 34–35) for a similar objection. John Gardner (2014: 12) offers criticism outside the context of war. He rightfully points to a number of legal officials who ought not to act by fidelity of deference: the "judges, the barristers, the solicitors, the court clerks, the prosecutors, and so on."

25 See Section 4.1 for more on the distinction between revisionist and traditionalist approaches to war.

26 Harel (2014: 104) offers two caveats to this position. One requirement that must be met for an individual to become a public official is that the public office must be "necessary for the performance of legitimate state functions" (which, for instance, a concentration camp guard would violate). The second is that the individual should only "defer to those orders that fall within the scope of the responsibilities of public office" (which, for instance, would preclude killing fellow citizens). Despite these caveats, soldiers will still be expected, on his account, to obey many types of unjust orders. Harel (2014: 105) acknowledges as much:

> It is often the case that it is permissible or even required for a public official to act in accordance with an order, although it is impermissible or unjust on the part of the state to issue such an order.

But, as I have argued, this is problematic. It is one thing to say that individuals ought not to be blamed for following orders that they did not perceive to be unjust. It is quite another to demand that they follow clearly unjust orders to uphold fidelity of deference.

27 For the difference between direct and indirect risks, see Section 3.3 and Sections 4.1–4.3.

28 We must remember that the collective authorizing entity includes *both* civilian legislators and high-ranking military officers who represent collective interests. See Section 5.4 for more on this point.

29 Dorfman and Harel (2013: 80) make a similar point about the underspecified nature of collective dictates.

30 This question, of course, is at the center of debates on liability to attack in war—on whose contributions to a war effort are sufficient to make them liable to attack and whose contributions are so minimal that they are entitled to non-combatant immunity. These important questions are beyond the scope of this book.

31 See Section 4.6 for more on why individuals must, on occasion, accept their own deaths to avoid exposing others to unjustifiable risk.

References

Avant, Deborah (2005). *The Market for Force: The Consequences of Privatizing Security* (Cambridge: Cambridge University Press).

Baker, Deane-Peter (2011a). *Just Warriors, Inc.: The Ethics of Privatized Force* (London: Continuum International Publishing).

Baker, Deane-Peter (2011b). "To Whom Does a Private Military Commander Owe Allegiance?," in Paolo Tripodi and Jessica Wolfendale (eds), *New Wars and New Soldiers: Military Ethics in the Contemporary World* (Surrey: Ashgate), pp. 181–198.

Daniels, Mitchell (2003). "Office of Management and Budget Circular A-76 (Revised)." Available at: www.whitehouse.gov/sites/default/files/omb/assets/about_omb/a76_incl_tech_correction.pdf (accessed December 28, 2015).

Dorfman, Avihay and Alon Harel (2013). "The Case Against Privatization," *Philosophy and Public Affairs*, Vol. 41, No. 1, pp. 67–102.

Fabre, Cécile (2007). "Mandatory Rescue Killings," *The Journal of Political Philosophy*, Vol. 15, No. 4, pp. 363–384.

Fabre, Cécile (2012). *Cosmopolitan War* (Oxford: Oxford University Press).

Fabre, Cécile (2014). "Rights, Justice, and War: A Reply," *Legal Theory*, Vol. 33, No. 3, pp. 391–425.

Feaver, Peter (1996). "The Civil–Military Problematique: Huntington, Janowitz, and the Question of Civilian Control," *Armed Forces & Society*, Vol. 23, No. 2, pp. 149–178.

Feaver, Peter (1998). "Crisis as Shirking: An Agency Theory Explanation of the Souring of American Civil–Military Relations," *Armed Forces & Society*, Vol. 24, No. 3, pp. 407–434.

Feaver, Peter (1999). "Civil–Military Relations," *American Review of Political Science*, Vol. 2, pp. 211–241.

Feaver, Peter (2003). *Armed Servants: Agency, Oversight, and Civil–Military Relations* (Cambridge, MA: Harvard University Press).

Gardner, John (2014). "The Evil of Privatization," paper presented at the University of Warwick, May 16, 2014. Available at http://users.ox.ac.uk/~lawf0081/pdfs/privatization.pdf (accessed December 28, 2015).

Hammes, T.X. (2010) "Private Contractors in Conflict Zones: The Good, the Bad, and the Strategic Impact," *National Defense University Strategic Forum*, No. 260. Available at http://psm.du.edu/media/documents/reports_and_stats/think_tanks/inss_hammes-private-contractors.pdf (accessed December 28, 2015).

Harel, Alon (2008). "Why Only the State May Inflict Criminal Sanctions: The Case Against Privately Inflicted Sanctions," *Legal Theory*, Vol. 14, No. 2, pp. 113–133.

Harel, Alon (2011). "Outsourcing Violence," *Law & Ethics of Human Rights*, Vol. 5, No. 2, pp. 395–413.

Harel, Alon (2014). *Why Law Matters* (Oxford: Oxford University Press).

Hedahl, Marcus (2009). "Blood and Blackwaters: A Call to Arms for the Profession of Arms," *Journal of Military Ethics*, Vol. 8, No. 1, pp. 19–33.

Heinecken, Lindy (2014). "Outsourcing Public Security: The Unforeseen Consequences for the Military Profession," *Armed Forces & Society*, Vol. 40, No. 4, pp. 625–646.

International Committee of the Red Cross (2009). *The Montreux Document*. Available at www.icrc.org/eng/assets/files/other/icrc_002_0996.pdf (accessed December 28, 2015).

Jinks, Roy G. and Sandra C. Krein (2006). *Images of America: Smith and Wesson* (Charleston: Arcadia Publishing).

Kasher, Asa (2008). "Interface Ethics: Military Forces and Private Military Companies," in Andrew Alexandra, Deane-Peter Baker, and Marina Caparini (eds), *Private Military and Security Companies: Ethics, Policies, and Civil–Military Relations*, (New York: Routledge), pp. 235–246.

Krahmann, Elke (2007). "Transitional States in Search of Support: Private Military Companies and Security Sector Reform," in Simon Chesterman and Chia Lehnardt (eds), *From Mercenaries to Market: The Rise and Regulation of Private Military Companies* (Oxford: Oxford University Press), pp. 94–112.

Krahmann, Elke (2008). "The New Model Soldier and Civil–Military Relations," in Andrew Alexandra, Deane-Peter Baker, and Marina Caparini (eds), *Private Military and Security Companies: Ethics, Policies, and Civil–Military Relations* (New York: Routledge), pp. 247–265.

O'Brien, Kevin (2007). "What Should and Should Not Be Regulated," in Simon Chesterman and Chia Lehnardt (eds), *From Mercenaries to Market: The Rise and Regulation of Private Military Companies* (Oxford: Oxford University Press), pp. 29–48.

Pattison, James (2014). *The Morality of Private War* (Oxford: Oxford University Press).

Singer, Peter Warren (2007). *Corporate Warriors: The Rise of the Privatized Military Industry*, updated edition (Ithaca: Cornell University Press).

Taussig-Rubbo, Mateo (2009). "Outsourcing Sacrifice: The Labor of Private Military Contractors," *Yale Journal of Law & the Humanities*, Vol. 21, No. 1, pp. 101–164.

United Kingdom Defence Forum (2015). "Fact Sheet – FS 56." Available at www.ukdf.org.uk/Fact_Sheet_-_FS_56.html (accessed May 9, 2015).

United Kingdom Ministry of Defence (2014). *UK Armed Forces Annual Personnel Report*. Available at www.gov.uk/government/uploads/system/uploads/attachment_data/file/312539/uk_af_annual_personnel_report_2014.pdf (accessed December 28, 2015).

United States Office of the Deputy Assistant Secretary of Defense (2013). *2013 Demographics: Profile of the Military Community*. Available at www.militaryonesource.mil/12038/MOS/Reports/2013-Demographics-Report.pdf (accessed December 28, 2015).

Wolfendale, Jessica (2008). "The Military and the Community: Comparing National Military Forces and Private Military Companies," in Andrew Alexandra, Deane-Peter Baker, and Marina Caparini (eds), *Private Military and Security Companies: Ethics, Policies, and Civil–Military Relations* (New York: Routledge), pp. 217–234.

Wulf, Herbert (2008). "Privatization of Security, International Interventions and the Democratic Control of Armed Forces," in Andrew Alexandra, Deane-Peter Baker, and Marina Caparini (eds), *Private Military and Security Companies: Ethics, Policies, and Civil–Military Relations* (New York: Routledge), pp. 191–202.

8 Challenges

Thus far, we have seen that privatization of the rank-and-file may be justifiable so long as high-ranking military officers and commanders are public officials (according to the governance and punishment arguments) and so long as risk-imposing agents in the field are placed under a uniform chain of command (according to the control argument). I turn now to four prominent critiques of military privatization, which target even restricted forms of the practice. I address these critiques to defend my theory of military supply against compelling alternatives—some with a long history in political theory—and to reinforce the right of collectives to privatize rank-and-file military functions in a limited fashion.

This chapter begins by discussing the distinction drawn by Pattison (2014: 8–12) between contingent and deeper objections (first introduced in Section 5.2). My book is concerned with the latter, but I canvass the former to dispel any worries that a set of potentially damaging challenges has been overlooked. The four sections that follow then take up and reject four deeper objections to military privatization. In confronting these deeper objections, I will strengthen the theory of military privatization that has been defended in this book: military force must be authorized publicly, as demanded by the risk-prevention and compensation arguments, but *may* be supplied privately or publicly so long as it satisfies the requirements of the governance, punishment, and control arguments.

8.1 Contingent objections

The difference between contingent and deeper objections, according to Pattison (2014: 8–12), is that contingent objections may be neutralized through proper institutional regulation, while deeper objections apply even in the presence of optimal institutional regulation.[1] Because contingent objections cannot easily decide against privatization in moral examinations of war like this one, my analysis has focused (and will continue to focus) on deeper objections. But a brief look at the range of potential contingent objections reveals how *current* practices deviate substantially from moral principles that ought to govern war.

Pattison masterfully catalogues these objections in a three-pronged framework: those directed at the *employees* who serve as contractors, those directed at the *employers* (e.g., states) who rely upon contractors, and those directed at the *international system* in which contractors operate. Each objection depends upon one or more empirical premises, which will not be probed here. For the sake of argument, we may assume each to be true. The key point that I want to bring out—which Pattison endorses—is that these objections, even if sound (as I believe them to be), would cease to apply in the presence of effective regulation and thus cannot undermine military privatization *as such*.[2]

Let's consider each contingent objection in turn, beginning with *employees* of the private security industry. The most oft-cited worry directed at contractors themselves is what Pattison (2014: 32) calls the *violation of human rights objection*.[3] Private actors are more likely than regular soldiers to violate human rights on the battlefield[4]—not because they are inherently depraved but because governments and PMSCs devote fewer resources to hire the right individuals and/or appropriately manage those who have been hired.[5]

The second category of contingent objections raised by Pattison concerns the *employers* of private contractors.[6] (1) *The argument from effectiveness* (Pattison 2014: 85–89): The employment of contractors may be financially wasteful because of poor competition for contracts, overcharging, fraud, and the drain of capable soldiers from the public to the private sector; military privatization can also compromise strategic success on the battlefield due to the perceived illegitimacy of contractors among local populations (making it hard to win "hearts and minds"), insufficient time for contractors to train together, the unwillingness of contractors to share information with one another freely given competition, and an unduly narrow military focus (rather than a broader strategic focus) in the execution of missions. (2) *The argument from democratic control* (Pattison 2014: 100–103): Governments may use contractors to circumvent democratic constraints and acquire political buffer for unpopular action; they may employ contractors with little transparency; governments can be subject to the influence of industry lobbyists; and, on the whole, they may fail to inculcate democratic values in private contractors as they do in regular military personnel. (3) The *argument from the treatment of employees* (Pattison 2014: 107–110): PMSCs may mistreat employees with little consequence, having allegedly committed human trafficking and labor violations in recent years, provided contractors with insufficient gear and poor medical care, suppressed information about contractor deaths, engaged in deceptive recruiting practices, and discriminated based upon gender.

Finally, Pattison highlights several worries that military contracting poses for the current *international* legal system. (1) *The formal constraints argument* (Pattison 2014: 144–149): International and domestic laws do not apply straightforwardly to contractors; the use of contractors can allow states to surreptitiously authorize war (e.g., proxy wars), and non-state actors may hire contractors to use military force in a way that flouts international law. (2) *The informal constraints argument* (Pattison 2014: 153–156): Because the private security industry lacks transparency, the international community cannot easily make normative judgments about states that employ contractors, thus diminishing the potential for such normative judgments to constrain states. (3) *The insecurity argument*

(Pattison 2014: 159–161): Only the well-off can afford to hire contractors, which fosters inequality between states and within states. (4) *The instability argument* (Pattison 2014: 166–169): The multiplication of private actors with a capacity to supply military force may increase the likelihood that wars will be fought in general and, in particular, that unjust and/or unilateral wars will be fought.

Each of the contingent objections raised in this section has a potential regulatory solution. Regarding employees, governments have a variety of mechanisms available to reduce adverse selection and moral hazard, both for contractors and regular military personnel (Baker 2011a: 85–143; 2011b). These mechanisms are likely to be particularly effective when rank-and-file personnel are placed under a uniform chain of command, as demanded by the control argument.[7] Regarding employers, the objections may be resolved through better financial oversight, incentives to retain soldiers in the regular military, heightened transparency, restrictions on lobbying, training to inculcate democratic values in contractors, and strengthened labor laws to punish PMSCs that mistreat employees. Regarding the international community, the formal constraints argument underscores the need for amendments to domestic and international law. The insecurity argument provides further reason to heighten transparency in international relations. The insecurity argument requires redistributive efforts to level the playing field for access to military protection. And the instability argument demands that governments curtail the scale of the private security industry and strengthen international control over the resort to force. All of these objections certainly challenge current contracting practices, but they do not challenge an optimally regulated system of military outsourcing.

8.2 Motivations

Having surveyed several important contingent objections to military privatization, let us turn to the first deeper objection: the so-called *motivationalist* objection.[8] According to this objection, hiring private actors is problematic, even when governments adopt restrictions of the sort I have defended (and even when they enshrine optimal regulations), because contractors possess objectionable motives. Tony Coady (2008: 15) argues that an individual "who hires his gun to the highest bidder or, less dramatically, fights predominately for money will typically lack the motive appropriate to war, as specified by just war theory." Pattison (2014: 37) gives a more precise rendering of this argument by defending three distinct premises: "(1) Motives matter in moral judgment. (2) It is problematic if individuals are predominately motivated by financial gain. (3) Private contractors are *more likely* to be predominately motivated by financial gain than regular soldiers."

While the first premise is sufficiently plausible to forego detailed analysis, the second premise, while also plausible (I will suggest), is less straightforward.[9] As Pattison highlights, two sorts of challenges are commonly raised against the second premise.[10] First, many individuals are motivated predominately by financial gain, including some doctors, lawyers, and so on, and surely we would not want to condemn *all* who are motivated predominately by financial gain (Baker 2011a: 38).[11] But it is not obvious that we *should* embrace these motivations in others spheres (Coady 2008: 219; Pattison 2014: 39). The fact that we tend not to condemn professionals who are motivated predominately by financial gain does not mean that we *ought to* endorse their behavior.

A second objection to the second premise acknowledges that financial motives may be objectionable but insists that these motives are less objectionable than patriotic zeal, which motivates many regular soldiers and can lead to dehumanization of the enemy (Lynch and Walsh 2000: 133–134; Coady 2008: 221; Baker 2011a: 37; Steinhoff 2011: 149; Fabre 2012: 221). Lynch and Walsh (2000: 146) underscore the potentially grave repercussions of unchecked patriotic zeal: "The terrible slaughter of the first two World Wars in this century were not, and *could not* have been, perpetrated by opposing mercenary forces." But the motivationalist argument does not endorse unchecked patriotic zeal as a motive appropriate to war; it simply requires that individuals be motivated predominately by *something* besides financial motives: "other-regarding concerns," whether for "family, community, state, humanity, or even fellow soldiers" (Pattison 2014: 40).

While the first two premises are sound, the third premise is more problematic. We should start by noting, as Pattison acknowledges, that some regular soldiers may *also* be motivated predominately by financial gain, particularly in militaries that promote career-building (Sandel 1998: 218; Lynch and Walsh 2000: 135–137; Percy 2007: 15; Coady 2008: 207; Krahmann 2008: 252; Wolfendale 2008: 218; Baker 2011a: 39; Steinhoff 2011: 139; Fabre 2012: 219). Moreover, contractors are not *always* motivated by financial gain and may instead possess other motives, including the national pride that drives many regular military personnel to serve (Lynch and Walsh 2000: 139–142; Percy 2007: 14–15; Coady 2008: 209, 223–224; Wolfendale 2008: 218; Taussig-Rubbo 2009: 140; Baker 2011a: 39; Steinhoff 2011: 139; Fabre 2012: 219).[12] Nonetheless, Pattison insists that the motivationalist argument is still a deeper objection, because "it could not be tackled by effective regulation given that motives are subjective ... and can be reasonably presumed to apply to private contractors to a greater extent than to regular soldiers" (Pattison 2014: 43).

But it seems that regulation *can* ensure that military contractors are no more likely than regular military personnel to possess improper motives. In order to challenge the third premise, we need not identify a set of regulations to ensure that *everybody* in the rank-and-file possesses motives appropriate for war (which would be impossibly difficult). We need only identify a set of regulations to ensure that contractors are not *more likely* than regular military personnel to possess improper motives. This could be accomplished through enhanced government oversight of PMSCs—in their hiring practices, training procedures, and termination policies. It might also be accomplished via improvements in the pay and benefits of regular military personnel. Such improvements would make the individual choice to work in the private sector less financially advantageous (relative to the military) and thus help guarantee that contractors are no more likely to be motivated by financial gain than regular soldiers.

More importantly, as Pattison (2014: 45) acknowledges, even if *all* military contractors possessed objectionable motives and these motives could not be excluded by enhanced oversight or improved remuneration in the regular military, *still* it is unlikely that these motives would be capable of rendering their service unjustifiable. The reason is that we would not want to punish those who would otherwise be protected by military contractors in war. Fabre (2012: 222) makes a similar point about individuals who become doctors for financial reasons:

> Even if it is wrong (*arguendo*) for someone to enter the medical profession mostly out of financial motives, and even if is wrong for a patient to hire him, surely the patient's interest in surviving is important enough to grant him a right to do so.

The same must be true of individuals who are being defended in war. If the war is just and has been collectively authorized, we would not want to penalize those who are being attacked by demanding that defenders have all of the right motives.

8.3 Chickenhawks

A second deeper problem has been advanced by Cheyney Ryan (2009). He dubs it the chickenhawk argument. According to Ryan, military privatization undermines a dictum that must govern warfare: "You should endorse a war only if you would be willing to fight in, even die in, the war yourself (or have your loved ones do so)" (Ryan 2009: 4). The validity of this argument depends upon the following: (*i*) the justifiable endorsement of public military authorization requires, *at a minimum*, that the individual proffering his or her endorsement be willing to participate in military supply; and (*ii*) one cannot fulfill the requirement of (*i*) through a willingness to participate in *private* supply (only in public supply). Thus, individuals who authorize war must be willing to participate in public supply. Moreover, (*iii*) individuals, and the states they constitute, who hire private entities to supply military force demonstrate an *un*willingness to participate in public supply. Thus, individuals, and the states they constitute, must not hire private entities to supply military force.

While (*i*) raises a number of questions, I wish to assume for the sake of argument that it is correct. It is (*ii*) and (*iii*) that are most problematic. According to (*ii*), individuals who endorse a war must sign up for regular military service rather than contracted service. Yet, an individual who decides to endorse a war and serve her country as a private contractor encounters all of the risks that regular military personnel encounter and contributes to all of the benefits to which they contribute. The chickenhawk argument does not seem capable of dictating the *kind* of military service that one who endorses a war ought to undertake. What is important for proponents of the chickenhawk argument is that individuals make *some* kind of military contribution to the wars that they endorse. Indeed, in at least one respect, proponents of the chickenhawk argument ought to favor military privatization. Because military contractors are capable of choosing individual wars for service—in contrast to regular military personnel who must typically serve in all wars—military contractors are able to participate in only the wars that they endorse (thus satisfying the chickenhawk dictum) while sitting out of wars that they oppose.

Yet, even if (*ii*) were correct and service in regular militaries was required, this does not mean that states must be prohibited from hiring private contractors—the claim of (*iii*). Suppose that everybody in a society who supported a given war served in the military to help prosecute that war. Why would the society then be prohibited from supplementing their war efforts with military contractors? The proponent of the chickenhawk dictum demands that *all* who support a war should fight, not that *only* those who support a war should fight. Given the problems of (*ii*) and (*iii*), the chickenhawk objection to military privatization must be rejected.

8.4 Markets

A third argument against military privatization, this one advanced by Michael Sandel (1998, 2009: 75–102), criticizes the role of markets in warfare.[13] Sandel argues that the infiltration of monetary exchange into certain human services, like military service, corrupts the values that these services aim to promote. Sandel distinguishes this argument from a different argument, which he labels the *argument from coercion*. The argument from coercion "points to the injustice that can arise when people buy and sell things under conditions of severe inequality or dire economic necessity" (Sandel 1998: 94).[14] Thus, many become military contractors because they have no other viable options for work and must support themselves and their families. Such arguments from coercion may be set aside as contingent; fair bargaining practices could be instituted to ensure that military contractors are not coerced into service.

The more threatening challenge, which is the one that Sandel defends, is the *argument from corruption*. This argument maintains that, regardless of the background conditions under which money is exchanged, still "certain moral and civic goods are diminished or corrupted if bought and sold for money" (Sandel 1998: 94). Hiring military contractors, on this account, would be problematic even if such contractors were wealthy individuals who enjoyed expansive job prospects and negotiated under fair conditions. The reason is that markets corrupt the good that military service aims to promote. Specifically, according to Sandel (1998: 108), markets corrupt the good of republican citizenship: "to be free is to participate in shaping the forces that govern the collective destiny," and only on the battlefield does one come to acquire the civic virtues that are needed for effective participation.

On this point, Sandel draws a helpful comparison with commercial surrogacy. He argues, following Elizabeth Anderson (1990), that the good of pregnancy is to promote an emotional connection between two intrinsically valuable beings, the child and mother. Commercial surrogacy violates this good by "requiring the surrogate mother to suppress whatever parental love she feels for the child" and thereby converting "women's labor into alienated labor" (Anderson 1990: 81). We might say that, just as the good of pregnancy promotes an emotional bond between the mother and her child, the good of military service promotes a communal bond between the collective and its citizens. By allowing markets to govern military service and individuals who serve to profit financially, the state severs the bond that would form on the battlefield between the citizens and the community for which they fight.

Another sphere of commercial activity that has been compared to private military contracting is prostitution (Sandel 1998: 95).[15] The argument for corruption applies similarly to prostitution as it does to commercial surrogacy. The good of sex, according to this argument, is to promote a bond between partners— in the way that pregnancy promotes a bond between the mother and child—and by allowing markets to govern sexual relationships, as it does in the case of prostitution, we contravene the good that sex is meant to promote. Private military contracting, on this analogy, functions like prostitution in that it severs the appropriate bond between the citizen and the state.

The difficulty with Sandel's objection to military privatization is threefold. First, it is not clear why citizens must serve in the military in order to develop the appropriate civic bonds to the state that are necessary for the protection of their freedom. The debate among republicans about the need for military service in republican polities stretches back to Machiavelli, who thought that military service was critical (Skinner 2002: 195–212).[16] But others have plausibly argued that what matters is not military service but rather some kind of *national* service (James 1970 [1910]; Dagger 2002: 7–12; 2005; Barber 2004: 298–303; Pattison 2014: 127). The key is that citizens do something "for other members of the community in return for the protection of the law and the other benefits that one hopes to continue to receive" (Dagger 2002: 5). Thus, the reliance upon private contractors by the state may not be worrying so long as citizens are able to do national service outside of the military.

Moreover, even assuming that military service is required, the two problems that confronted the chickenhawk argument are again relevant here. First, individuals could maintain strong civic bonds to the state by serving in war as private contractors. Certainly, the bond that is built between military contractors and the state may be weaker than the bond that is built between regular military soldiers and the state. But it seems that a contractor who risks life and limb to support a military effort would grow attached to that effort and to the authorizing entities in whose name he serves. Second, even if republican bonds could only be preserved through military service as a regular soldier (rather than private contractor), it is not clear that supplements to this national military service in the form of private contractors would be inconsistent with the good of republican citizenship. The state would not sever the bond between the citizen and the state; it might simply enhance the bond of others to the state.

8.5 Communal ties

A fourth and final deeper objection maintains that communal bonds are important but not for the reasons that civic-republicans like Sandel espouse. Pattison (2014: 110–111), for example, argues that, with military privatization, the "potential benefits in terms of communal identity that come from having citizens (either as a conscript or a volunteer) defend the community are generally lost"; as with other familiar community-building practices like voluntary blood donation, military service "strengthens communal ties and fosters a general sense of concern for other members of the community." But, as Pattison's comparison to voluntary blood donation indicates, a community might achieve strengthened identity in a number of ways. The question is: why might it be particularly important to achieve via the military?

The most plausible answer, it seems, is that strengthened collective identity is helpful for the prosecution of just wars. In this vein, Jessica Wolfendale (2008: 223–229) argues that the symbolism of national defense—the war memorials, parades, and so on—enhance public dialogue about war by focusing attention on those who suffer. Lindy Heinecken (2014: 637–638) suggests that strengthened collective identity may translate into better recruitment and greater self-sacrifice on the battlefield.[17]

But contractors may also promote communal bonds. As Pattison (2014: 111) acknowledges, contractors can protect vulnerable populations from attack and thus enable individuals to form communal bonds with one another (by keeping them alive). Moreover, contractors may promote communal bonds directly through their service and sacrifice. Taussig-Rubo (2009: 122) points to a memorial rock garden at one Blackwater compound, which is dedicated "to the courage and honor of our fellow teammates." Contractors, including five Fijians who died in Iraq, have received the highest civilian honor bestowed by the US government: the Defense of Freedom Medal (Taussig-Rubbo 2009: 150). As Spearin (2008: 209–210) writes: "US officials have praised the work of private US personnel, they have publicly mourned their deaths, and they have attended their funerals back in the United States." The notion that contractors are incapable of promoting communal bonds does not seem quite right.

More importantly, and these points notwithstanding, even if contractors do not promote communal bonds like regular soldiers, this does not undermine the *right* of states to hire private contractors. Perhaps states that hire private contractors may be less cohesive than states that rely upon national militaries. But any diminishment of communal bonds that results from the employment of military contractors in a limited fashion would not seem to keep states from prosecuting wars in a just manner. Indeed, Lynch and Walsh remind us that the communal bonds that *do* facilitate effective and just fighting tend to be bonds between military personnel rather than between the soldier and citizen. As they write, "[o]ne is committed to the fellows in one's patrol or unit or company or army— that is, those whom one fights *with* rather than fights *for*" (Lynch and Walsh 2000: 139). Because these bonds in the military may develop among private contractors and regular military personnel alike, many of the benefits that accrue through communal bonds with regard to the justice of wars may also accrue in a system with military privatization.

8.6 Conclusion

In this chapter, I have considered four deeper objections to military privatization, stemming from worries about the financial motivations of contractors, the chickenhawk behavior of collectives who hire contractors, the infiltration of markets into military service, and the diminishment of communal bonds that may result from privatization. None of these arguments succeeded in undermining the *right* of collectives to rely upon contractors in war.

Notes

1 Note that, on Pattison's (2014: 9) account, "for an objection to be fully deep it would need to apply to *all* PMSCs [or contractors], *not* apply to any regular soldiers or armed forces, and not be able to be feasibly tackled by regulation."
2 Needless to say, we have a moral duty to bring about such sound regulation.
3 The names assigned to each category of objections and the points raised under each are from Pattison (2014). I am simply summarizing Pattison's work on contingent objections in this section. Note that some of his contingent objections fall into more than one category—for example, objections directed at employers and the international system. For the sake of brevity, I include each objection only once.
4 This objection is a counter-point to Machiavelli's (2007 [1515]: 51) famous critique that mercenaries are not sufficiently ruthless: instead, they do "their utmost to keep themselves and their soldiers out of the way of fatigue and danger." The difficulty with this position is that Machiavelli was either wrong, in which case the view must be rejected, or he was right, in which case the gentility of mercenaries would speak *in their favor* (Baker 2008: 33; Coady 2008: 213; Lynch and Walsh 2000: 143).
5 The objection is perhaps more in keeping with a second contingent objection that Machiavelli (2007 [1515]: 48–49) offers against mercenaries: that they cannot be trusted. But Machiavelli's point is less compelling than objections from adverse selection or moral hazard, because mercenaries (and contractors more generally) are not *inherently* untrustworthy. As Coady (2008: 213) puts it, "in the contemporary world, the difference between mercenaries and other successful military leaders with regard to the temptation to abuse their office for political purposes may normally be insignificant." Baker (2011a: 137) takes this point one step further: "given that their motives tend to be pecuniary rather than political, it may even be thought that there is less of a danger of this sort of behavior from private warriors." See Section 8.2 for more on the motivations of private actors.
6 Note that some of Pattison's deeper objections have contingent components as well. We saw in Section 7.2 that the "problem of private choice" underpinned deeper objections based upon military efficacy, democracy, and distributive justice. All three of these deeper objections have associated contingent objections, which are presented below. The two other deeper objections raised by Pattison, based upon motivations and communal ties, are presented and challenged in Sections 8.2 and 8.5.
7 We should keep in mind, however, that the control argument is not itself a contingent objection. The chain of command is needed even in the presence of optimal regulation. It is needed to provide ongoing contact with rank-and-file personnel so that they have access to collective justifications for their risk-imposing behavior.

8 For this term "motivationalist objection," see Lynch and Walsh (2000: 135).
9 Uwe Steinhoff (2011: 141–142) has argued, *contra* the first premise, that motives do not matter in moral judgments about war, because individuals cannot control their motives. They can only control the actions that follow from those motives. Thus, so long as soldiers participate in just wars and *act* justifiably, they do not commit wrongs despite possessing improper motives. A complete discussion of Steinhoff's position is beyond the scope of this chapter. But his position is a substantial departure from the moral reasoning that has underpinned attributions of responsibility in this book and that typically underpin attributions of responsibility in moral philosophy and criminal law (Pattison 2014: 37).
10 Pattison (2014: 40) mentions a third challenge—that the premise is too demanding and does not allow individuals to properly pursue their own self-interests. But the motivationalist argument, on Pattison's formulation, objects only to individuals who are motivated *predominately* by financial gain, not to those who are motivated *at all* by financial gain.
11 In tackling what she calls the "profiteering objection"—the idea that "individual mercenaries live off the suffering others"—Fabre (2012: 224–226) likewise argues that many professionals, including doctors and lawyers, live off the suffering others (so to speak). Steinhoff (2011: 140) puts the argument in these provocative terms: "Physicians require diseases and benefit from them—does that mean that they cast aside their moral attitude towards diseases?." The answer, it seems, is no. Walzer (2006: 27) offers a similar argument about regular soldiers.
12 The similarities between the motives of regular soldiers and the motives of contractors are particularly striking in the case of contractors who have already served as regular military personnel. Philipps and Brennan (2015) quote an American Iraqi war veteran who went back to Iraq as a private actor to fight with the Pesh Merga against the Islamic State of Iraq and Syria (ISIS): "I may not be enlisted anymore, but I'm still a warrior."
13 Sandel's (1998) earlier work offers more philosophical argumentation for positions that are also endorsed in his later work (Sandel 2009: 75–102). I will quote the former except when material is present exclusively in the latter.
14 See Satz (2010: 5) for a similar distinction.
15 Civic-republicanism is not the only ground that has been employed for drawing an analogy between prostitution and private military contracting. An alternative ground concerns *objectification*—that those who purchase the services of prostitutes and private contractors objectify individuals who provide these services. In considering this challenge, Fabre (2012: 233) cites Kant (1991 [1795]: 95) on a similar point about regular soldiers:

> the hiring of men to kill or be killed seems to mean using them as mere machines and instruments in the hands of someone else (the state), which cannot easily be reconciled with the rights of man in one's own person.

But, as Fabre (2012: 224) argues, this position demonstrates the need for states to exercise due care when employing soldiers and contractors; it does not seem to demonstrate that relying upon soldiers and contractors is *necessarily* problematic.
16 Sandel (1998: 112) also cites Rousseau (1973 [1762]: 265): "As soon as public service ceases to be the chief business of the citizens and they would rather serve with their money than with their persons, the state is not far from its fall."
17 Pattison (2010: 438) cites Wolfendale in his earlier work in support of this argument and Heinecken in his later work (Pattison 2014: 111).

References

Anderson, Elizabeth (1990). "Is Women's Labor a Commodity?," *Philosophy and Public Affairs*, Vol. 19, No. 1, pp. 71–92.

Baker, Deane-Peter (2008). "Of 'Mercenaries' and Prostitutes: Can Private Warriors Be Ethical," in Andrew Alexandra, Deane-Peter Baker, and Marina Caparini (eds), *Private Military and Security Companies: Ethics, Policies, and Civil–Military Relations*, (New York: Routledge), pp. 30–42.

Baker, Deane-Peter (2011a). *Just Warriors, Inc.: The Ethics of Privatized Force* (London: Continuum International Publishing).

Baker, Deane-Peter (2011b). "To Whom Does a Private Military Commander Owe Allegiance?," in Paolo Tripodi and Jessica Wolfendale (eds), *New Wars and New Soldiers: Military Ethics in the Contemporary World* (Surrey: Ashgate), pp. 181–198.

Barber, Benjamin (2004). *Strong Democracy: Participatory Politics for a New Age*, Twentieth Anniversary Edition (Berkeley: University of California Press).

Coady, C.A.J. (2008). *Morality and Political Violence* (Cambridge: Cambridge University Press).

Dagger, Richard (2002). "Republican Virtue, Liberal Freedom, and the Problem of Civic Service." Paper presented at the Annual Meeting of the American Political Science Association, Boston, MA, September 2002.

Dagger, Richard (2005). "A Duty to Serve? Responsibility, Reciprocity, and the Paradox of Civic Service," *Public Policy Research*, Vol. 12, No. 1, pp. 15–21.

Fabre, Cécile (2012). *Cosmopolitan War* (Oxford: Oxford University Press).

Heinecken, Lindy (2014). "Outsourcing Public Security: The Unforeseen Consequences for the Military Profession," *Armed Forces & Society*, Vol. 40, No. 4, pp. 625–646.

James, William (1970 [1910]). "The Moral Equivalents of War," in Richard Wassertstrom (ed), *War and Morality* (Belmont: Wadsworth).

Kant, Immanuel (1991 [1795]). "Perpetual Peace: A Philosophical Sketch," in H.S. Reiss (ed), *Kant: Political Writings* (Cambridge: Cambridge University Press).

Krahmann, Elke (2008). "The New Model Soldier and Civil–Military Relations," in Andrew Alexandra, Deane-Peter Baker, and Marina Caparini (eds), *Private Military and Security Companies: Ethics, Policies, and Civil–Military Relations* (New York: Routledge), pp. 247–265.

Lynch, Tony and A.J. Walsh (2000). "The Good Mercenary?," *The Journal of Political Philosophy*, Vol. 8, No. 2, pp. 133–153.

Machiavelli, Nicolò (2007 [1515]). *The Prince*, in Peter Constantine (ed/trns), *The Essential Writings of Machiavelli* (New York: Random House).

Pattison, James (2010). "Deeper Objections to the Privatisation of Military Force," *The Journal of Political Philosophy*, Vol. 18, No. 4, pp. 425–447.

Pattison, James (2014). *The Morality of Private War* (Oxford: Oxford University Press).

Percy, Sarah (2007). "Morality and Regulation," in Simon Chesterman and Chia Lehnardt (eds), *From Mercenaries to Market: The Rise and Regulation of Private Military Companies* (Oxford: Oxford University Press), pp. 11–28.

Philipps, Dave and Thomas James Brennan (2015). "Unsettled at Home, Veterans Volunteer to Fight ISIS," *New York Times*, March 11, 2015.

Rousseau, Jean-Jacques (1973 [1762]). *The Social Contract*, translated by G.D.H. Cole (London: J.M. Dent & Sons).

Ryan, Cheyney (2009). *The Chickenhawk Syndrome: War, Sacrifice, and Personal Responsibility* (Lanham: Rowman and Littlefield).

Sandel, Michael (1998). "Commodification, Commercialization, and Privatization," *What Money Can't Buy: The Moral Limits of* Markets, The Tanner Lectures on Human Value. Paper presented at Brasenose College, Oxford, May 11–12.

Sandel, Michael (2009). *Justice: What's the Right Thing to Do?* (New York: Farrar, Strauss, and Giroux).

Satz, Debra M. (2010). *Why Some Things Should Not Be for Sale: The Moral Limits of Markets* (Oxford: Oxford University Press).

Skinner, Quentin (2002). *Visions of Politics, Volume II: Renaissance Virtues* (Cambridge: Cambridge University Press).

Spearin, Christopher (2008). "Privatized Peace? Assessing the Interplay Between States, Humanitarians, and Private Security Companies," in Andrew Alexandra, Deane-Peter Baker, and Marina Caparini (eds), *Private Military and Security Companies: Ethics, Policies, and Civil–Military Relations* (New York: Routledge), pp. 203–216.

Steinhoff, Uwe (2011). "Ethics and Mercenaries," in Paolo Tripodi and Jessica Wolfendale (eds), *New Wars and New Soldiers: Military Ethics in the Contemporary World* (Surrey: Ashgate), pp. 137–151.

Taussig-Rubbo, Mateo (2009). "Outsourcing Sacrifice: The Labor of Private Military Contractors," *Yale Journal of Law & the Humanities*, Vol. 21, No. 1, pp. 101–164.

Walzer, Michael (2006). *Just and Unjust Wars: A Moral Argument with Historical Illustrations*, 4th edition (New York: Basic Books).

Wolfendale, Jessica (2008). "The Military and the Community: Comparing National Military Forces and Private Military Companies," in Andrew Alexandra, Deane-Peter Baker, and Marina Caparini (eds), *Private Military and Security Companies: Ethics, Policies, and Civil–Military Relations* (New York: Routledge), pp. 217–234.

9 Conclusion

This book has provided a comprehensive theory of military privatization. Part I asked questions about the authorization of war, and part II asked questions about the supply of war. Five major arguments were given, the first two related to military authorization (Part I) and the final three related to military supply (Part II).

1 *The risk-prevention argument*: Public actors have a right, and indeed a responsibility, to prevent private actors from authorizing military force (*contra* A_1 and A_2).

   P_1 *All Affected Fundamental Interests*: A decision ought to be withdrawn from the private sector and reserved for public discharge when one course of action under consideration imposes risk, above some threshold level, to the fundamental interests (e.g., physical security) of enough individuals. Fundamental interests are interests that are sufficiently weighty so as to be protected by rights.

   P_2 *Risk-Imposition of War*: The private decision to authorize military force imposes risk (directly and indirectly), above an acceptable threshold level, to the fundamental interests of enough individuals. Specifically, the private decision to authorize military force imposes such risks to our fundamental interests in physical security.

2 *The compensation argument*: Public monopolizing entities have a responsibility to authorize military force when the security of those who have been disarmed is threatened (*contra* A_0).

3 *The governance argument*: Governments must not outsource the responsibilities carried out by high-ranking military officers to the private sector (*contra* A_3).

   Defense from Political Power:

   P_1 If the decisions made by some set of individuals must be publicly discharged, then the responsibilities of those who exercise sufficient political power over such decisions must also be publicly discharged.

P_2 High-ranking military personnel exercise sufficient political power over public civilian decisions of military authorization and military supply for us to insist that the responsibilities of these individuals must be publicly discharged.

Defense from Military Decision-Making:

P_3 If the decisions made by one set of individuals must be publicly discharged, then the decisions made by a second set of individuals must also be publicly discharged if the two sets of decisions are sufficiently similar in some morally decisive sense.

P_4 The decisions of high-ranking military personnel in military supply *are* sufficiently similar to the public decisions of civilians in military authorization in a morally decisive sense. The morally decisive sense in which they are similar is that both have the potential to cause extensive destruction on the scale that wars are fought.

4   *The punishment argument*: Governments must not outsource the responsibilities carried out by commanders to the private sector (*contra* A_3).

P_1 *Discipline Premise*: Military commanders must be able to discipline soldiers who disobey orders (or who, in some other way, transgress) by imposing severe constraints on the soldiers' freedom.

P_2 *Penal Authority Premise*: The severe constraints on the soldiers' freedom that are sometimes required to preserve military discipline are not constraints that a private agent may justifiably impose. Only public agents should have the moral authority to impose such constraints.

5   *The control argument*: Militaries must include risk-imposing agents under a uniform chain of command.

P_1 *Unauthorized risk-imposition*: If a member of the rank-and-file imposes risks that have *not* been collectively authorized by relevant risk-bearers, then that individual imposes risks upon individuals who lack a say in her decision of risk-imposition, which is unjustifiable.

P_2 *Transmitted justification*: Only a uniform chain of command allows members of the rank-and-file to impose collectively authorized risks.

These five arguments establish the central role of public actors in the authorization and supply of military force.

If the conditions of these five arguments are satisfied, *then* a government may be justified in privatizing rank-and-file military functions. The justification for this right is that collectives have an obligation to protect those who have been disarmed (according to the compensation argument), and, to do this, they require a capacity to purchase (or produce) weapons and employ individuals to prosecute collectively authorized wars. Once the discretion of contractors has been limited (according to the control argument), any morally relevant difference between contractors and regular military personnel dissolves. Collectives may thus fulfill their responsibilities of military authorization via reliance upon either contractors or regular military personnel to serve in rank-and-file positions. But the sphere carved out for permissible outsourcing is narrow.

With the arguments of this book now in place, let us return to the concept with which we started: the Middleman State. Defined by small bureaucracies and ambitious policy agendas, the Middleman State ensures that its citizens are provided with goods: military protection, education, health care, domestic policing, garbage collection, space exploration, and many others. But the Middleman State does not *itself* furnish these goods. Instead, it relies upon private contractors for their supply.

Recall the mantra of the Middleman State: "the business of government is not to provide services, but to see to it that [services] are provided" (United States House Committee on Government Reform 2006: 1).[1] In some sense, this book has affirmed the core commitment of the Middleman State in matters of war. When our basic rights are threatened, what is most important, in the end, is that *somebody* fights just and collectively authorized wars on our behalf and that *somebody* thereby defends us.

Yet, this book also represents a strong challenge to the Middleman mantra. I have insisted that, in order for a government to properly see to it that services are provided in war, the government must provide *some* of these services itself. In some cases, the business of government *is* to provide services. Most fundamentally, governments must retain the choice over which wars to fight. Furthermore, once wars have been chosen, governments must not outsource the responsibilities of high-ranking military officers or commanders. Only by preserving the public character of the civilian legislature and the military leadership and by instituting a robust chain of command can a government ensure that the service in question—war—is being properly provided. Middleman practices are thus permissible only assuming the *public* performance of these governmental functions.

But so long as governments retain authority over war—as specified by the risk-prevention argument (Chapters 3 and 4), the compensation argument (Chapter 4), the governance argument (Chapter 5), the punishment argument (Chapter 6), and the control argument (Chapter 7)—Middleman practices cannot be ruled out. Collectives must not be denied the right to protect the fundamental interests of those who are threatened.

Note

1 This quote, we may recall, is from Mitch Daniels, former head of the US Office of Management and Budget.

Reference

United States House Committee on Government Reform (2006). *Dollars Not Sense: Government Contracting Under the Bush Administration* (Washington, DC: Government Printing Office).

Index

Page numbers in *italics* denote tables.

Fabre, Cécile *continued*
121n16; defense of private military
supply 100, 102–5, 121n16, 150, 152–3,
163; on legitimate authority 38n3, 39n8,
123n31; on mercenaries 101, 121n11;
and risk-prevention argument 45–6,
51–2, 63n21, 63n22, 78, 89, 91, 92
Fainaru, Steve 2
Feaver, Peter 122n25, 153
fidelity of deference 157–9, 164, 166n24,
166n26
fidelity of reason 158
Fidell, Eugene R. 132
financial motivation 171–2
Finland 92
Finlay, Christopher J. 15n25, 26, 27, 38n3,
39n11, 41n31, 41n33, 63n21, 93n3, 144n5
Foot, Phillipa 94n19
force-short-of-war 6, 28, 34, 39n18, 75
formal constraints argument 170, 171
Forsyth, Frederick, *Dogs of War* 3
France, summary punishment 132
Frank, Thomas 1, 17n37
free market economy of military force 10, *11*
friendly fire 160
Frisk, Daniel 14n6
fundamental interests 12, 54, 55–60,
63n27, 71–4, 77–8, 79, 83, 85, 87, 88,
91, 94n12, 100, 126–7, 159, 160, 163;
see also all affected fundamental
interests premise

Gardner, John 166n24
Gat, Azar 75
Gates, Robert 130
Gaynor, James 132
genocidal wars 5, 15n15
Ginsberg, Wendy 2
Glanz, James 2
global justice 85–6
Goodin, Robert 52–3, 63n23
governance argument 13, 100, 105–9,
119–20, 181; defense from military
decision-making 109, 116–19, 120, 182;
defense from political power 109–16,
120, 181–2
Gross, Michael L. 27
guerrilla movements 4, 6, 32, 105
Gulf War 2, 3, 14n6
Guttman, Israel 88

Hamilton, Alexander 140
Hammes, T.X. 130, 151
Handfield, Toby 15n18

Hannsson, Sven Ove 94n9
Hardin, Russell 122n23
Harel, Alon 15n20, 16n26, 145n25, 150,
157–9, 163, 164, 165n4, 166n22,
166n23, 166n26, 167n29
Hart, H.L.A. 63n28
Hedahl, Marcus 130, 144n8, 151
heightened probability claim 67–8, 73–6
Heinecken, Lindy 130, 151, 176
Held, Virginia 15n25, 25, 38n3, 48, 86,
144n5
Hersey, John 83
Hersh, Seymour 144n7
Hessians 6, 15n21
high-ranking military personnel 13, 14,
100, 105–7, 109, 122n23, 144n4, 160,
181–2; decision-making *in bello*
116–19; political power of (and
decisions of continuation 113–6; and
military authorization 111–13, 115)
Hohfeld, Wesley 26
humanitarian intervention 10, 45, 90, 92–3,
93n5, 102; bystanders 50, 51, 63n19,
63n21; consent of intended beneficiaries
(victims) 50, 51, 62n14, 62n15, 62n16,
63n19; representation of intended
beneficiaries 50; and risk-prevention
argument 49–52
Huntington, Samuel 122n25

imprisonment 131, 132, 133, 135, 136,
137, 138–9, 140
indirect-risk imposition 51–2, 67–8;
alternative participation objection 76,
77–8, 81; heightened probability claim
67–8, 73–6; moral relevance claim
67–8, 76–81; precautionary principle
83–4; prohibited-response objection 76,
78–80, 81; participation in decisions of
military authorization 76–81
individual-claim 29–30, 34
informal constraints argument 170
innocent bystanders 35–6, 79–80, 89–90,
94n8
insecurity argument 170–1
instability argument 171
interest theory of rights 55, 63–4n28
International Commission on Intervention
and State Sovereignty (ICISS) 75
international institutions, strengthened 85,
86, 87
international law 69, 74, 170, 171
internecine killings 135
interstate militarized disputes 75